AWA PRESS

HOW TO GAZE AT THE SOUTHERN STARS

'Will beguile experienced star-spotters and absolute beginners alike'

Maggie McDonald, *New Scientist*

'An intimate and elegant guide to the splendours above'

Stephen James O'Meara, *Sky and Telescope*

'A wonderful read that should spark enthusiasm for the splendours of the night sky'

Anna McIntyre, *The Marlborough Express*

'Richard Hall can only be described as astronomy's answer to David Bellamy'

Tanya Katterns, *The Dominion Post*

'Don't measure the success of this book by how often you pick it up, but by how often you put it down while you head outside to find what Richard Hall is talking about. You'll be doing that a fair bit'

David Hill, *New Zealand Books*

'An idiosyncratic mix of astronomy, starlore and anthropology ... bound to inspire the reader to cast an eye heavenwards'

Marilyn Head, *The Weekend Herald*

'An entertaining ride through the night skies'

Linda George, *The Wellingtonian*

'Hall's enthusiasm is infectious'

Denis Welch, *Listener*

'Informative and stylish. Hall is an excellent communicator'

Robin List, *The Manawatu Evening Standard*

02

THE GINGER SERIES

OTHER TITLES IN THE GINGER SERIES

01 How to Watch a Game of Rugby
Spiro Zavos

03 How to Listen to Pop Music
Nick Bollinger

04 How to Pick a Winner
Mary Mountier

05 How to Drink a Glass of Wine
John Saker

06 How to Catch a Fish
Kevin Ireland

07 How to Look at a Painting
Justin Paton

08 How to Read a Book
Kelly Ana Morey

09 How to Catch a Game of Cricket
Harry Ricketts

how to gaze at the southern stars

richard hall

AWA PRESS

First edition published in 2004 by
Awa Press, 16 Walter Street,
Wellington, New Zealand

Reprinted 2005, 2006

National Library of New Zealand Cataloguing-in-Publication Data
Hall, Richard.
How to gaze at the southern stars / by Richard Hall. 1st ed.
(The ginger series: 2)
ISBN 0-9582509-9-5
1. Southern sky (Astronomy)—Popular works. I. Title II. Series.
520—dc 22
ISSN 1176-8452

Printed by Astra Print, Wellington
Printed on environmentally friendly and chlorine-free Munken paper.
This book is typeset in Walbaum

www.awapress.com

For Margaret and Rick Hall, my parents,
who first inspired me to question and explore
the universe of which we are all a part

ALSO BY RICHARD HALL

The Work of the Gods (with Kay Leather)

ABOUT THE AUTHOR

RICHARD HALL is not only one of New Zealand's leading astronomers but an outstanding communicator, with a gift for making complex scientific information understandable and enjoyable for the ordinary person. His 2004 Radio New Zealand series *Summer Stars* reached and inspired thousands. He also has a monthly astronomy programme on Newstalk ZB Wellington. Hall is senior public programmes officer at Wellington's Carter Observatory, and a founding member of Phoenix Astronomical Society. His latest project, Stonehenge Aotearoa, is a unique Pacific-inspired adaptation of the original Stonehenge – especially designed for its location in rural Wairarapa.

CONTENTS

ILLUSTRATIONS

The known is finite,
the unknown infinite;
intellectually we
stand on an islet in
the midst of an
illimitable ocean of
inexplicability.
Our business in every
generation is to reclaim
a little more land.

T. H. Huxley,
1887

Why gaze at the stars?

THE HUMAN FASCINATION with outer space is reinforced in a song most New Zealanders learned as children:

Twinkle, twinkle, little star,
How I wonder what you are.
Up above the world so high,
Like a diamond in the sky.
Twinkle, twinkle, little star,
How I wonder what you are.

I started to wonder at a very young age and this book draws on my lifelong passion for all things astronomical.

I hope to share some of my awe of the cosmos with those of you about to begin what has been for me a fascinating, exciting and unending journey. Astronomy is such a vast subject, with links to every other science, that no book, however large, can do it justice. However, I hope this book will stimulate you, the reader, to discover for yourself some of the amazing objects the universe contains.

My interest in the stars began when I was at junior school. I lived about an hour's train ride from London. My mother often took my sisters and me to the great museums in London. My favourite was the Natural History Museum, and the most inspiring place the fossil gallery. When I first walked into the enormous gallery and was confronted with the skeleton of a diplodocus, a 97-foot long dinosaur that walked the Earth 120 million years ago, I was hooked. Once upon a time there really were dragons! I became fascinated with the Earth's past – the pageant of life and the ever-changing geography and environment of our world.

My mother also took us to the movies regularly and one evening she took us to see *Invaders from Mars.* In this film a young boy sees a flying saucer landing in a field during a thunderstorm. No one in his town believes him, and the Martians, hidden in their subterranean spaceship, begin to control the inhabitants, including the boy's own father and mother.

Watching this was a terrifying experience, but it also set me thinking. That night when we came home the sky was clear and studded with stars. I can remember wondering if one of those twinkling points of light was Mars. Perhaps there were strange beings out there, across the depths of space? When I considered the marvels that had occurred on this world, I began to contemplate what might exist elsewhere. That wonderment remains with me to this day.

We live in a universe that is built on a scale beyond the comprehension of the human mind. There are more stars in the known universe than there are grains of sand on all the beaches of the world. Each star is a sun. In all likelihood, orbiting around each of those suns is a system of planets.

The number of worlds in the known universe must be almost countless. Just about anything we can imagine, and more, probably exists out there somewhere. It is this, the grand mystery of the universe, plus its exquisite beauty, that captivates me. For me astronomy is an adventure of the imagination into time and space.

A lot of people think the rest of the universe is somehow remote, and of no great importance to our daily lives. Nothing could be further from the truth. Our star – the sun – provides us with warmth and light without which life on Earth would be impossible. Its energy drives our weather systems; fluctuations in this energy can produce droughts, floods and ice ages. The moon, and to a lesser extent the sun, causes the tides to ebb and flow. Giant meteors occasionally strike the Earth, devastating the environment and changing the course of life. Cosmic radiation causes genetic changes in living things. We are intimately connected to the rest of the universe – and it's a lot closer than you think. As the famous English astronomer Fred Hoyle once remarked, 'Space isn't remote at all. It's only an hour's drive away if your car could go straight upwards.'

I have often been asked if the vastness of the universe makes me feel insignificant. It doesn't. Human beings are not something separate from the rest of the universe; we are part of it. The atoms and molecules that make up our bodies were manufactured in the interiors of stars millions of years before the Earth was born. When these stars died they hurled their substance out into space. This material was eventually swept up in the formation of new stars and planets, one of which was our world. Each of us is literally made of stardust. Raised to a level of consciousness, we are the universe looking at itself.

How to get started

WHAT DO YOU need to become an astronomer? For a lot of people astronomy, along with many other sciences, lives in the too-hard basket. On more than one occasion someone has told me he or she has always been interested in astronomy but doesn't know enough maths to become involved. But you don't need to be a mathematician to gain an understanding of astronomy, any more than you need a degree in botany to appreciate and learn about a native forest. All you need is enthusiasm.

Unless you intend to remain an armchair astronomer, you will first need to familiarise yourself with the night sky. You do not need a telescope to take up stargazing; all you need is your eyes. Use the charts in this book to find your way among the stars, planets and constellations. When I started all I had were simple star charts and a red torch to read them by. The latter can be an ordinary small torch covered with red cellophane.

Why do astronomers use red lights? If you go outside on a clear dark night from a brightly lit room you will find you can see little. At first you will see only the brighter stars. Then, slowly, fainter and fainter stars become visible. It takes at least ten minutes for your eyes to become fully adjusted to the dark. By this time the magnificent intricate structure of the Milky Way will be discernible. Should you now use a bright white light you will instantly lose your 'dark adaptation'. Red light affects the eyes' dark adaptation the least, so if you need to look at charts use a red light that is just bright enough to read by.

The only other must is warm clothing; even on a summer's night you can get cold if you are not moving around. The winter night sky in New Zealand is magnificent, but to enjoy it you will need thermal underwear, a woollen or polypropylene hat, woollen socks and gloves.

A telescope is not essential. Before you consider getting one I suggest you invest in a pair of binoculars. A good pair of binoculars will show you a lot more than a cheap telescope. They are a worthwhile investment because you will still use them even if you later acquire a telescope – and of course they are useful for purposes other than stargazing.

For astronomy, the most important characteristic of a telescope or binoculars is the aperture, the diameter of the objective lens or primary mirror. This is what gathers the light and determines how faint an object you can see. The naked eye is limited in what it can see by the aperture of its iris, which is at best about 5mm in diameter. A telescope with a 150mm aperture lens effectively gives you an eyeball the size of a cartwheel.

Two numbers, for example 6x40, designate the primary characteristics of a pair of binoculars. The '6x' is the magnification and the '40' the aperture in millimetres. Although magnification is not as important as the aperture, it must be borne in mind that the higher the magnification the smaller the field of view. Because the best views of star fields and the Milky Way are achieved with a large field of view, I recommend a low magnification for binoculars. In addition, if the magnification is much more than 8x you will have difficulty holding the binoculars steady without extra support. The bigger the aperture the better, but bigger aperture means a bigger price. I consider the best all-round binoculars for astronomy are 7x50. These are what the navy uses for night vision.

If you get hooked on astronomy, it probably won't be long before you want to acquire a telescope. A good astronomical telescope is expensive; if you are buying new, you are unlikely to get much change out of a thousand dollars for even a modest instrument. On the market there are a lot of cheap telescopes whose manufacturers exaggerate what they are capable of, and if you are not careful you will be investing in a junky toy. Furthermore, no single telescope design is suitable for all aspects of astronomy. Long focal-length telescopes are good for observing the planets, while short focal-length telescopes are best for extended objects such as galaxies (island universes of billions of stars held together by gravitational attraction) and nebulae (large interstellar clouds of gas and dust).

Some people are disappointed when, looking through their $250 telescope, they can't see what the Hubble Space Telescope can. But when you think about it, why would NASA spend over two billion dollars on the Hubble if all they had to do was buy a plastic fantastic from a department store? My suggestion is that you join a local astronomical society and try out different telescopes before you purchase your own. There you can often get good second-hand telescopes, plus good advice from experienced observers. The astronomical society to which I belong has available for use by members a range of large telescopes which would be beyond the financial reach of most people.

This is not a textbook on astronomy. It is an introduction to some of the magnificent wonders of our southern night sky, and the colourful myths and legends associated with the stars. It is a fireside book that can be taken outside to help you discover the stars for yourself. I hope it will encourage you to try.

First I discuss the origins of star lore, and why astronomy was a cornerstone of the rise of civilisation. Next I explain our Earth-bound view of the night sky, and 'how it all works'. Then we take a tour of the night sky, identifying bright stars, important southern constellations and some of the wonders of the universe. Bear in mind that any tour of the night sky can never be comprehensive, simply because of the sheer scale of the universe. I will introduce you to stars and constellations that are easy to identify, but be aware there is much, much more.

Throughout I discuss both the science and the myths associated with celestial objects. Science reveals the wonders of the universe, while the myths dramatically show how our ancestors attempted to gain an understanding of the universe around them.

Two possibilities
exist: either
we are alone in
the universe
or we are not.
Both are equally
terrifying.

Arthur C. Clarke

Ancient campfires

HUMAN FASCINATION with the stars began long ago. Imagine if you will the red glow of sunset fading over the African savannah. As night falls a small family of our ancestors huddles around the campfire. Robbed of vision, they are vulnerable in the darkness. Throughout the night they hear the call of wild animals, the lion, leopard and hyena. The night is the time of the predator. The flames of the fire comfort them: as well as providing warmth and light, they help keep the predators at bay.

As they listen to the crackle of the fire and the sounds of the night they look upwards and watch the stars.

What, they wonder, are these mysterious lights in the sky? Some of the stars flicker, just like their campfire. Perhaps they are the distant campfires of other wanderers? The bright stars, and the patterns they formed, would have been very familiar to these early stargazers. While their world was full of uncertainty, the stars had a permanence and predictability that must have offered some reassurance.

More than five thousand generations separate us from this small family gathering where, by the campfire, a journey began that would change the way of life of the human species. About 100,000 years ago our ancestors emerged from the cradle of Africa and migrated into Asia. About 60,000 years ago they reached what is now the continent of Australia. These migrants occupied tropical or subtropical regions where seasonal changes were minor and food supplies abundant. Although they used fire and built shelters, they manufactured only simple hand-held stone tools. Living in small family groups, their total numbers were small and they had little impact on the physical environment.

About 40,000 years ago – 60,000 years after the diaspora from Africa began – our species underwent its first population explosion. A race of people with a new and advanced technology emerged in Europe and Asia Minor. As well as rapidly overrunning territories occupied by the first migrants, they moved into temperate and arctic regions. By the end of the ice age eight to ten thousand years ago, they had inhabited most major landmasses. By a thousand years ago all habitable landmasses on Earth had been reached, Aotearoa-New Zealand being the last.

What generated this great expansion? Why were these people so successful? Unlike their predecessors, the new people fashioned finely crafted tools and weapons. They used needles and thread to make garments and footwear, and hunted with bows and arrows and harpoons. They also created works of art, and traded goods over widely separated areas. They were still hunter-gatherers living in small nomadic groups, but there had been a huge leap in technology and culture.

The new culture originated in a temperate zone with marked seasonal changes. To survive, these people had to cope with extremes of climate, and a food supply that varied markedly with the seasons. This required extensive advancement in technology, which in turn demanded a large increase in the knowledge base of the group. Complex information needed to be passed from person to person, and from generation to generation. We know that the forebears of this new people, unlike those before them, hunted large and dangerous game, an activity requiring planning, teamwork and rapid communication. It seems highly probable that what made this all possible, and these people different from their predecessors, was the emergence and development of a complex language.

Many animals, particularly primates, which live in social groups use a simple language to communicate with each other – a number of different sounds or gestures which are equivalent to words. With the possible exception of dolphins, humans are unique in having the ability to string words together to convey complex information.

Language, the ability to both formulate complex ideas *and* articulate them to others, may have resulted from genetic changes within one particular isolated group of people. It is sobering to think that every society in the world, every culture and every language may have a common origin in a small band of men and women who lived 40,000 years ago – a group of people of whom we know nothing.

Following the stars

FOR THOSE PEOPLE who migrated beyond the African Eden, knowledge of the stars was essential. Like their more recent counterparts on the North American plains, they would have followed the seasonal migration routes of large game animals. The ability to navigate would have been vital, particularly for people who dwelt on vast open terrains and had to travel great distances, and they undoubtedly used the sun, moon and stars.

Most of the names of stars in common use today are Arabic in origin. Arabian people were great navigators, across oceans not of water but of sand. Imagine being confronted with a searing hot, featureless desert. Out there somewhere, beyond the horizon, is an oasis you will need to find if you are to survive. In the vast expanse of drifting sands there are no roads or landmarks. How do you find your way? Like the people who lived 40,000 years ago, you use the stars.

For a nomad living in the northern latitudes it was vital to be able to predict seasonal changes. Game animals, when they migrate, move much faster than people do on foot. You wake up one morning and the forest is silent – the animals and birds have moved on. If there are no berries or vegetables your food source has vanished overnight. Arrive at a place too early and there may be no food to gather; leave too late and rising rivers or falling snow could trap you. Knowing when to move on was often the difference between life and death. But how would you know?

Each season different stars are seen in the night sky. However the same stars appear at the same time each year. Our ancestors would soon have realised that the appearance of certain stars heralded coming seasonal changes. They started linking these stars together to form patterns they could easily recognise. Today we call such star patterns 'constellations'. The names given to important stars or constellations often reflected something that happened in that particular season. It could be the appearance of certain animals or birds such as the swan (Cygnus), the lion (Leo) or the bear (Ursa Major). Or it could be a seasonal change such as a shift in winds or coming of rains.

On the subject of constellations, even astronomers get confused. I read in one book that they originated from 'peoples who fancied that they could see a likeness to certain fabled creatures and mythological heroes among the stars'. This is wrong.

In the planetarium at the Carter Observatory I have often been asked by visitors to pick out a particular constellation. It is usually a constellation of the zodiac – the swathe of the sky along the celestial paths of the sun, moon and planets. And the most common reason for requesting the particular star group is that it is the person's star sign.

When the constellation is pointed out, the person is often surprised (and disappointed) to find the pattern of stars has little or no resemblance to the constellation's name – the name of their star sign. This is because constellations were named for their symbolism or practical significance, not because they had a likeness to mythological creatures. For example, the constellation of Aquarius (The Water Carrier) got its name not because it looks like a person pouring water from an urn, but because when these stars rose just before the sun it signified the onset of the rainy season. The water carrier was a symbol of the gods pouring water down upon the Earth. What was important about these stars was their meaning, not the pattern they formed.

Of course what was important to one group of people might not be as important for another group living in a different climatic zone. Consequently, cultures that were widely separated geographically had different meanings for the same stars or constellations. The symbolism of some of the more important constellations, including those of the zodiac, originated so long ago that it is common to cultures throughout the world. From place to place and time to time the names may be different, but the figure of stars and their meaning is much the same.

Other ancient constellations, however, are unique to a particular culture, and have a special meaning relating to the way of life, locality and environment of those people. It is important to remember that the stars of a constellation often have no actual physical relationship with each other; it is people who have created the connection.

The names of the constellations in modern use have been agreed upon by a body called the International Astronomical Union. The entire sky has been divided into 88 specific regions and these are now *the* constellations.

Most of the names given to these constellations are of classical Greek origin, and in most cases they incorporate the stars of the ancient constellations of the same name. However people in different countries still use unofficial names from their folklore: these are called asterisms. Examples are 'the Pot' (Southern Hemisphere), 'the Sickle' (Northern Hemisphere) and 'the Plough' (Europe). The Plough, the stars around the North Pole, are as important to people in the Northern Hemisphere as the Southern Cross is to people in the Southern. The Americans call this constellation 'the Big Dipper'.

The stars that form an asterism may be part of a larger constellation, or they may be a combination of stars from more than one constellation. Examples of this are the False Cross, a combination of four stars very similar to the Southern Cross, and the Summer Triangle, a group of three bright stars which lies across the Milky Way.

I'm a poor underdog

But tonight I will bark

With the great Overdog

That romps through the dark.

Robert Frost,
'Canis Major'

Why the dog is a star

THE STORY OF the Dog Stars illustrates how star names can contain veiled meanings. When I was a child I often spent school holidays in the countryside with my grandmother. Being well away from city lights, the night sky was like black velvet strewn with stars. I still remember one particular warm autumn evening when my father had come to visit. As night fell he and I sat outside, watching a myriad of stars appear. One star outshone all the others, a piercing white point of light that flickered and flashed like a distant bonfire. My father said it was called the Dog Star. Later I discovered the name Dog Star was common knowledge, but no one seemed to know why the star was so called. Many decades later, when I studied the mythology of stars, I discovered the answer.

The Dog Star is the folk name for Sirius, the brightest star in the entire sky. Many people assume Sirius is called the Dog Star because it is the brightest star in the constellation of Canis Major, the Greater Dog. But in fact the name Canis Major originally applied only to the star: the constellation was built up later. The same is also true of the Lesser Dog constellation, Canis Minor. In ancient times only the constellation's brightest star was called Canis Minor. The Greeks called it Antecanis, or Northern Sirius. Today we call this star Procyon.

Sirius and Procyon sit on either side of the Milky Way. In ancient Egypt, Sirius was associated with the great goddess Isis: the star was said to be her immortal spirit. Isis married her brother Osiris, god of the Nile and the afterworld, and the name Sirius may be derived from his name. Sirius began appearing on Egyptian monuments and temple walls and being worshipped throughout the Nile Valley from 3285 BC.

The ancient Egyptians believed the Earth to be a reflection of the immortal heavens, with the Milky Way the celestial River Nile. Sirius's rising before the sun in mid-summer marked the beginning of the inundation of the Nile valley, and the start of Egypt's new year. The flooding of the Nile valley was the most significant event of the year for the Egyptians. It brought fertility to the

1 The Dog Stars at dawn 5,000 years ago.

land, which was fundamental to the survival and prosperity of the entire community. Because the Egyptians believed all events on Earth were controlled by gods, they attributed the flooding of the Nile to Sirius – hence the association with Isis, goddess of fertility.

Now to answer the question – why are Sirius and Procyon called the Dog Stars? From earliest times shepherds have used dogs to guard their flocks and warn of approaching danger. While the Nile flooding was essential for the Egyptians' survival, it also represented a physical danger to people and animals living on the flood plain. Sirius and Procyon were the shepherd dogs in the sky, standing guard on either side of the celestial Nile. Their mid-summer rising warned shepherds that the flood was coming and it was time to move the flocks to higher ground.

With the passage of time the meaning of the Dog Stars changed. Sirius means 'the scorching one' and in 500 BC its rising before the sun corresponded with the sun's entrance into the constellation of Leo. This marked the hottest time of the year, hence the term 'dog days': a dog day afternoon is the hottest part of a hot and hazy summer day.

In Europe and the Middle East this hot period was an unhealthy time of year, and the rising of Sirius was associated with disease. Homer wrote of the Dog Star:

The brightest he, but sign to mortal man
Of evil augury

And Pope:

Terrific glory! For his burning breath
Taints the red air with fevers, plagues and death

Good fortune or bad, the dog gets the blame!

Within this story of the Dog Stars are names and sayings which are in common use today, but most of us have no idea of their origin and meaning. The same is true of the names of the other stars and constellations. Each, though, has a fascinating story to tell.

Lost in the twilight

ANYONE WHO IS familiar with the night sky will know that individual stars or constellations are visible for several months. Furthermore, each night a given star or constellation will appear to rise or set a little earlier than the night before. How, then, did people in ancient civilisations read the stars in order to predict seasonal changes?

If you watch the stars from evening to evening, you will notice they are slowly moving westward. Eventually a given star becomes too close to the sun to be seen and is lost in the western twilight. About a month later the star reappears, rising in the east just before the sun.

This cycle, from one dawn rising to the next, takes exactly one year. The ancient Egyptians saw it as a cosmic cycle of birth, death and resurrection. When the star disappeared in the western evening twilight they believed it had been consumed by the fires of the sun. Its reappearance in the eastern dawn was seen as a rebirth from the same fires.

This appearance, when the star rises for the first time just before the sun, is the precise meaning of the term 'heliacal rising'. In most cultures this 'rebirth' was seen as a portent of a seasonal change or event. For example, to Maori the first rising of the bright star Whanui, commonly known as Vega, towards the end of February was a sign the kumara crop had matured and summer would soon be drawing to a close.

It was almost certainly the observations of such apparent relationships between the 'rebirth' of a star or constellation and an important seasonal event that led to the founding of the pseudo-science of astrology – the belief that events on Earth are controlled by the stars, and that the future can be discovered by people who know how to 'read' the stars.

The ancient Greeks were sceptical about this, and the rise of modern astronomy from the time of the Renaissance showed the tenets of astrology were based on misconceptions. For example, the fact the first pre-dawn rising of Sirius occurred just before the flooding of the Nile was merely a coincidence. These days due to the effects of precession – a 26,000-year cyclic wobble in the Earth's axis – these two events no longer coincide. Nonetheless these early misconceptions have been carried through space and time to every corner of the world.

Astrology and astronomy are often confused; more than once I have been erroneously described as an astrologer. The significant difference, in my view, is that astrology is a belief system – that is, it relies on faith, rather than science.

New age pseudo-science may appear on the surface to be quite logical and have some foundation in fact. However, anyone with even rudimentary scientific knowledge can easily detect misconceptions and fallacies. For example, the full moon is variously held responsible for lunacy, crimes and miscarriages, and that's just for starters. The pseudo-science goes something like this: The moon influences the Earth as it raises tides. Human beings are made mostly of water, so it stands to reason that the moon must also have an effect on us.

This sounds all very logical. However, the tidal effect of the moon is due to the difference in the gravitational tug between one side of the Earth and the other, a span of 12,576 kilometres. The difference in the gravitational pull of the moon over the length of a human body, at best two metres, is miniscule. A ride in an elevator will have a greater effect on you than the moon moving overhead. And riding an elevator does not cause madness, at least not in the normal course of events.

We are accustomed to think of myths as the opposite of science. But in fact they are a central part of it: the part that decides its significance in our lives.

Mary Midgley, ***The Myths We Live By***

The storyteller

TODAY KNOWLEDGE and information are readily available. Sit back and imagine how you would cope if there were no clocks, calendars, books, telephones, radios, television sets and computers. Our present society would collapse into total chaos. Yet for most of human history there were no such things as books or writing, let alone clocks and electronic devices.

How was information stored before the invention of the written word? In many communities a small number of carefully selected individuals were trained to remember specific knowledge such as genealogy, star lore and history of the tribe. Training began at an early age and continued through to adulthood. These individuals became living books, the 'wise' men or women who could be called upon to provide the community with accurate information.

Because information had to be remembered and recited accurately, it was incorporated into stories, poetry and song. One example is the famous nursery rhyme:

Ring-a-ring a roses
A pocketful of posies
Atishoo! Atishoo!
We all fall down.

I learnt this rhyme long before I went to school. My friends and I would join hands in a circle and dance around singing it. When we came to 'We all fall down' that's just what we did, usually in gales of laughter. Was this nothing more than a fun game for children – or does it have an original meaning, now forgotten?

In fact the rhyme originated in seventeenth century England, when the Black Death, or bubonic plague, was sweeping across Europe killing millions. Every line contained important information a child at that time needed to know. 'Ring-a-ring a roses' were the red circular swellings that appeared on the face and body of someone who had the disease. 'A pocketful of posies' indicated that flowers and herbs should be carried to ward off the plague. 'Atishoo! Atishoo!' was a reminder that sneezing was the first symptom of infection, and 'We all fall down' that those who caught the plague would die.

I was astonished when I found out that, as a child, I had been acting out the symptoms of the Black Death. Probably most schoolteachers also had no idea of the nursery rhyme's chilling origin. But what the example demonstrates is that placing in story form important information that people need to easily and accurate recall is an effective strategy. If you remember the story, you remember the information.

I have come to believe that all ancient stories contain information important for survival. Astronomy is rich in mythology, but the tales are often trivialised by people, particularly scientists, who think they are simple stories containing no useful information. My research has revealed that, to the contrary, these stories contain a wealth of meaning and knowledge that is still relevant today. In fact, I would argue that the very foundations of civilisation are built on the knowledge contained within ancient star lore. In this sense there is no such thing as mythology.

Human beings,
vegetables, or
cosmic dust,
we all dance to a
mysterious tune,
intoned in the
distance by an
invisible player.

Albert Einstein

Rise of the great traditions

DURING THE LAST ice age the nomadic way of life began to give way to permanent settlements based on horticulture. The oldest permanent settlements – at least the ones we know about – are found in the Jordan Valley and date back 19,400 years. Settlements of a similar age have been found in Japan. As the ice retreated and the climate warmed 15,000 to 8,000 years ago, large numbers of settlements began to appear in fertile areas such as the great river valleys.

In time, scattered communities within the same geographical area became linked by trade routes. The first of these appeared in Mesopotamia (modern-day Iraq) and central Africa about 12,000 years ago.

Perhaps due to their location on these trade routes, some settlements became marketplaces, and from these arose the first towns. Jericho, built at an oasis in the Jordan Valley, dates back some 9,000 years, making it perhaps the oldest city in the world.

These trading centres became powerful economic and political units and eventually evolved into the first city-states – walled cities which controlled the land directly around them. Similar centres of settlement began to appear independently elsewhere: in the Indus Valley 8,500 years ago, China 8,000 years ago and Central America 7,000 years ago.

Because each of these great centres emerged in isolation from the others, each had its own distinct culture and belief system. With the exception of hunter-gatherer peoples who have remained completely isolated, every culture in the world can trace the roots of its language, values and belief system back to one of these early centres of settlement.

As these centres evolved, new social orders emerged, with well-defined class structures, and division and specialisation of labour. Star lore, particularly its astrological or future-telling aspects, became the exclusive knowledge of a rising priesthood. These priests predicted awesome celestial events such as solar and lunar eclipses. They could foretell the changing of winds, the flooding of rivers and the coming of rains. It must have seemed to the ordinary person that they were communicating directly with the gods.

Interestingly, every major religion in the world is based on star lore. If this surprises you, think of the Cosmic Dance of Shiva – the Hindu depiction of the creation and death of the universe; the Black Stone which is Islam's most venerated object, kissed and touched by millions of pilgrims at the annual Hajj – actually a meteorite believed to have been discovered by Abraham (the Saudi Arabian city of Mecca was built where it fell); and the Star of Bethlehem – which I believe was a conflation of two celestial events, the conjunctions of bright planets and a nova, or exploding star. The Magi or Wise Men were, by the way, professional astrologers and probably came from Babylon.

Far from being of passing interest, then, astronomy was central to the lives of our ancestors, and inextricably linked to the rise of civilisation. It provided the human species with timekeeping, around which we order our lives; navigation, which allowed us to explore the planet; and the ability to predict and take advantage of seasonal changes. And the symbolism of stars and their planets became integral to spiritual beliefs. All this came about from observations, deductions and theories of people about whom we know little or nothing, not even their names.

Astronomy compels the soul to look upwards and leads us from this world to another.

Plato
(*circa* 427–347 BC)

Travels in space

EVERYONE KNOWS what a star is. Or do we? We talk of the stars 'coming out' at night but in fact a star, the sun, dominates the daytime. Our planet, Earth, orbits around it. The sun is a typical star: it appears much bigger and brighter than the other stars only because of its comparative closeness.

Because we can't judge the distance of celestial objects by sight alone, few of us appreciate the sheer size of the sun. The distances of all but the nearest of celestial objects are measured in millions of kilometres, numbers so large they are beyond the comprehension of the human mind. Some people think astronomers have a special ability that somehow allows us to comprehend the vastness of the universe. Unfortunately not. Just like you, when I get beyond the number seven I have to start counting.

The way I imagine astronomical distances is by scaling them down to something I'm familiar with. For example, the circumference of the Earth at its equator is 40,076.594 kilometres. We have all driven in a car at 100 kilometres an hour. If you could travel in a straight line at 100 kilometres an hour without stopping, you could drive all the way around the Earth in just 17 days. Yet just two and a half centuries ago it took Captain James Cook three years to circumnavigate the world.

If it would take just 17 days to travel around the Earth, how long would it take to drive to the moon? The answer is 160 days, or more than 5 months. This tells you that the moon is more distant than you probably imagined. Pictures or diagrams of the Earth and moon give the impression they are a lot closer than they really are. In fact the moon's average distance from the Earth is 384,400 kilometres, or 30 times the diameter of the Earth.

If it would take five months to drive to the moon, how long would it take to drive to the sun? The answer is you couldn't – you would die of old age before you got there. Travel time to the sun, at 100 kilometres an hour, would be 171 years. The sun's distance from the Earth varies by about 3 million kilometres over the course of a year, but the average is just under 150 million kilometres, making it 390 times further away than the moon.

This average distance between the Earth and the sun is called the 'astronomical unit' and is used to measure the scale of the solar system. The principal reason for Captain Cook's first voyage to the Pacific was to observe a transit of Venus – a rare situation where the planet Venus crosses directly between the Earth and the sun; his observations would then enable scientists to accurately measure the astronomical unit.

The fact the sun is so far away and yet appears so large in our sky shows its physical size must be enormous. The orb of the sun – the ball we see in the sky, excluding all the surrounding atmospheres – is so huge it could contain over 1.3 million Earths. This star of ours is a gigantic sphere of white-hot gas, mostly hydrogen and helium. It is far too hot for liquids or solids to exist.

The sun's substance is a little different to the gases on Earth. Due to its high temperature – 6,000 degrees Celsius at the surface, rising to 14 million degrees at the core – electrons are stripped from the atomic nuclei, producing a soup of electrified particles. This material, called plasma, still behaves like a gas but it can be compressed to enormous densities. Near the centre of the sun it is denser than lead.

In ancient times the sun was worshipped as a god. This makes good sense because, although it's not a sentient being, it is the source of almost all the energy on Earth, and our very existence depends upon it. Without the heat and light of the sun, temperatures on Earth would fall to almost absolute zero. Not only would the oceans freeze but so would the air we breathe.

Almost all the energy we use comes from the sun. We acquire energy by eating plants, or animals that eat plants. The energy stored in plants comes from sunlight. The wood or coal we burn is stored solar energy. The electricity we use in our homes and businesses comes from coal-fired generators or from hydro-generation. The water that fills the dams and drives the generators comes from rain evaporated from the oceans by solar energy. All energy chains lead back to the sun: we live entirely on sunlight.

In one second the sun releases more energy than the human race has used in the last 10,000 years. This enormous flood of energy comes from the core, where hydrogen is being converted into helium by a thermonuclear process similar to that used in a hydrogen bomb, but on a much larger scale. Every second of its existence the sun converts 600 million tonnes of hydrogen into 596 million tonnes of helium. In this nuclear reaction, 4 million tonnes of matter per second are annihilated and converted into energy. It is this that powers the sun – and our planet.

The sun is therefore a diminishing asset. Every second of its existence it becomes 4 million tonnes lighter. It has been doing this for billions of years. Fortunately its mass is so vast it will be able to continue to do it for billions of years to come.

This, then, is a star – a huge sphere of glowing plasma that generates its own heat and light. Planets, by comparison, are minor bodies that accompany a star. They are essentially debris left over from the formation of the star. They have little heat and no light of their own, and shine only by reflecting the light of their parent star. All the planets we can see in our night sky – and of course Earth itself – belong to the sun. They are the children of the sun, the remnants of the material from which the sun formed.

The stars we see in the night sky are distant suns, and each probably has its own planetary system. Some are physically similar to our sun; others are very different. The galactic zoo is rich in star species. The distances of these stars from us, and from each other, are so great that astronomers have had to abandon miles and kilometres and devise new units to measure the distances. The most common is the light-year.

In a vacuum, light travels at just under 300,000 kilometres a second. Nothing in the universe can move faster than light. If you could travel at the speed of light, you could travel right around the Earth in 0.13 of a second – literally in the blink of an eye. And the imaginary journey to the moon that took five months by car would be reduced to 1.3 seconds.

Because light travels at a constant speed, instead of saying the moon is 384,400 kilometres away we could say it is 1.3 light-seconds away. Our imaginary journey to the sun that would have taken 171 years in a car would take only 500 seconds, or 8.3 minutes. Therefore, the sun is 8.3 light-minutes away. But these are our near neighbours. The other stars are so remote their distances are measured not in light-minutes but in light-years. A light-year, then, is the distance light travels in a year, or 9.46 million million kilometres.

This brings us to the subject of time. Because light from the sun takes eight minutes to reach the Earth, we see the sun where it was, and as it was, eight minutes ago. Our world is so small relative to the speed of light (0.04 light-seconds in diameter) that, as far as time is concerned, there is only now. The past is gone and the future is yet to occur. But when we look into space the distances of celestial objects are so great we are looking into the past. And the further we look into space, the further back in time we see. When I look at Neptune, the most distant of the outer planets orbiting the sun, through a telescope I see it as it was four hours ago.

For stars the time gap is even more dramatic. The bright star Alpha Centauri – one of the Pointers to the Southern Cross – is, at 4.4 light-years, the nearest star to Earth. We see it as it was more than four years ago. Wezen, a star in the constellation of Canis Major, the Great Dog, is one of the most distant stars able to be seen without a telescope. The light that falls into our eyes when we look at this star started on its journey towards us at about the time of the birth of Christ.

Sometimes I am asked how astronomers know what was happening in the universe billions of years ago. It's easy: if you want to know what the universe was like a billion years ago, you look a billion light-years into space. All this is, of course, very problematic for the galactic empires depicted in movies such as *Star Wars* and *Star Trek*. If there were a rebellion on some world around a distant star, it would be ancient history by the time the news reached the emperor.

Had I been present at the Creation, I would have given some useful hints for the better ordering of the universe.

Alphonso the Wise
(*circa* 1270)

The celestial sphere

AS THE EARTH turns on its axis, the sun, moon and stars slowly move across the vault of the sky. Consequently, the appearance of the night sky changes throughout the night. In addition, the Earth is orbiting around the sun, so the constellations you can see change from season to season. The tilt of the Earth's axis, coupled with its orbital motion around the sun, causes the sun, moon and planets to move in ever-changing and sometimes complex paths across the sky. At first this may seem very confusing, but in fact everything is changing in a very ordered and predictable manner.

If you sit for a while under the stars you will soon become aware that the sky overhead appears to be moving. Stars that started out close to the western horizon will have disappeared, and new stars will have appeared in the east. We talk of the sun, moon and stars rising and setting but in reality the stars are motionless; it is we who are moving. Here in Aotearoa-New Zealand, due to the Earth's rotation we are moving at about a thousand kilometres an hour eastward. As the Earth turns from west to east, celestial objects appear to move in the opposite direction – from east to west.

Looking at the night sky you can appreciate why our ancestors imagined the stars were attached to an invisible sphere. The stars are so distant that our sense of perspective fails us, and all celestial objects appear to be at the same distance, namely infinity. To understand our Earth-based view of the universe and how we map the sky, it is useful to use the ancient concept of a celestial sphere.

In figure 2 we see the Earth rotating on its axis within a stationary sphere of fixed stars. We feel no motion as the Earth rotates because everything around us, including the air, is moving with us, at the same speed. It appears to us that it is the star sphere that is moving. The Earth rotates a full 360 degrees every 24 hours, so the sun, moon and stars move across the sky at the rate of 15 degrees per hour. Because their motion is due to the Earth's rotation, our view of the celestial sphere, and

2 The celestial sphere.

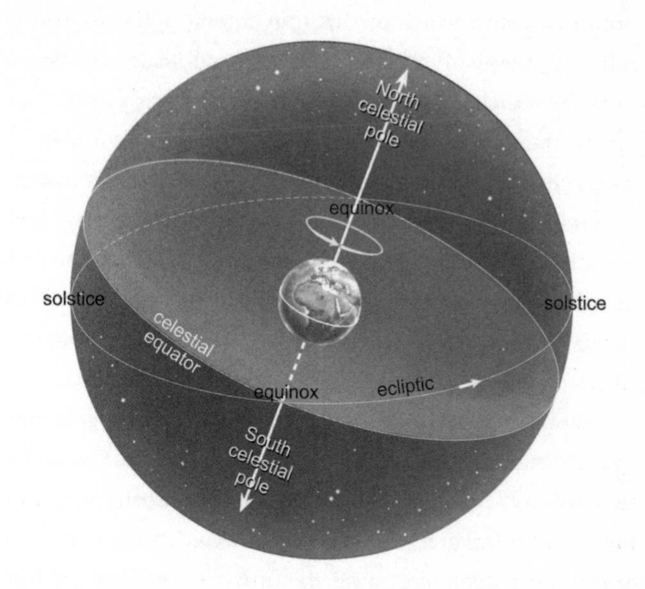

the way in which it appears to move, will depend upon our position on the Earth.

To understand how it works, we start by drawing an imaginary line that runs through each pole of the Earth (the line of axis about which the Earth rotates) and then extend this line outward at either end to the celestial sphere. The entire sky will appear to rotate around the points where this line touches the celestial sphere. These two points are called the celestial poles.

We can also project the Earth's equator out to the celestial sphere. This great circle marks the celestial equator. The equatorial line is exactly 90 degrees from each pole. Stars south of this line belong to the southern celestial hemisphere and are referred to as southern stars. Those to the north are called northern stars.

In figure 3, observer A is standing at the South Pole. As the Earth rotates the stars will appear to move slowly in circles around the celestial pole. Seen from here, the stars never rise or set. The celestial equator is at the horizon so our observer can only see southern stars. The reverse would apply if the person were standing at the North Pole.

Observer B is standing at the equator. The celestial equator runs directly overhead from east to west, and the two poles are located at the horizon due north and due south. The stars move in great circles parallel to the celestial equator, rising in the east and setting in the west. Our observer sees the stars of both hemispheres.

Most people live at a latitude somewhere between these two extremes. Observer C is standing at latitude 45 degrees south, midway between the South Pole and equator. The celestial pole is midway between the zenith (the point directly overhead) and the horizon, 45 degrees above the horizon looking due south. Stars within 45 degrees of the south celestial pole never set. Those within 45 degrees of the north celestial pole never rise.

We must now add another imaginary line or circle to our celestial sphere. This one is similar to the lines of longitude on Earth. It runs from the celestial pole,

3 The apparent motion of the stars as seen from different latitudes: (A) the South Pole, (B) the equator, and (C) 45 degrees south.

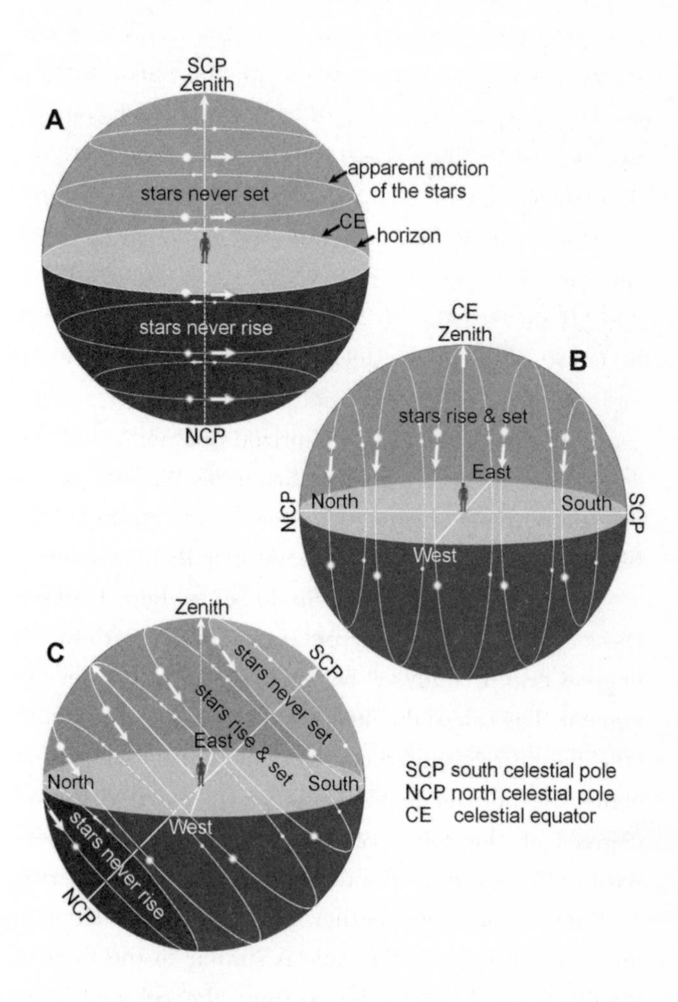

passing directly overhead through the zenith, across the celestial equator, and then down to the horizon (which is due north in the Southern Hemisphere), and finally onwards to the other celestial pole. This line is known as the meridian. When a star crosses the meridian it is said to culminate – that is, it reaches its highest point in the sky for that latitude. When the sun crosses the meridian it is noon.

At latitude 45 degrees a star that is 45 degrees from the pole will touch the horizon at its lowest point, but will not actually set. At its highest point it will cross the meridian directly overhead. Stars that pass directly overhead are called zenith stars.

By this time you may have noticed that the observed altitude of the celestial pole is directly related to your latitude on Earth. This is fundamental knowledge for the navigator. The celestial pole is always at a number of degrees above the horizon equal to the latitude at which you are standing. If you are standing at the South Pole, latitude 90 degrees south, it will be directly overhead, 90 degrees above the horizon. At the equator, 0 degrees latitude, the celestial poles are at the horizon, 0 degrees above the horizon. Wellington, New Zealand, for example, is located at 41 degrees south. From Wellington the south celestial pole is 41 degrees above the horizon.

If you can locate the celestial pole you learn two things. First, you establish your compass bearings, because the south celestial pole is directly above the horizon looking due south. Once you have established south, the other cardinal points (east, west and north) follow. Secondly, if you measure the altitude of the celestial pole you can determine your latitude. First and foremost, however, you need to be able to identify the celestial pole, and to do this you must be able to recognise important circumpolar stars. I will show you how to do this in the section called 'The Southern Cross' (page 113).

The fault, dear
Brutus, is not in
our stars,
But in ourselves,
that we are
underlings.

William Shakespeare,
Julius Caesar

The Age of Aquarius

AS THE EARTH orbits around the sun our night-time backdrop of stars slowly changes. Each evening a star will rise or set four minutes earlier than it did the night before. For example, an evening star that rises at 9 p.m. tonight will rise at 8.56 p.m. tomorrow, 8.52 p.m. the next evening, and so on.

We do, however, see the same stars at the same time each year. Stars that are opposite the sun in January will, during that month, be in our sky all night. Six months later the same stars cannot be seen, because they will be close to the sun and in our daytime sky. Come January they will be in our night sky again. Because the seasons are directly related to the Earth's position in its orbit, we see different stars at different seasons.

The solar system – the sun and its system of planets – is shaped like a large flat disk, with the sun at the centre and the planets orbiting within the disk. The Earth and the other planets orbit around the sun's equator, and in the same direction as the sun rotates on its axis. Seen from the sun's north pole, the planets would seem to orbit in an anti-clockwise direction. This is the same direction in which the Earth rotates on its axis, and the same direction in which the moon orbits the Earth.

Due to the orbital motion of the Earth, the sun appears to move against the background stars. This apparent path of the sun is called the ecliptic, and the background stars along this path form the constellations of the zodiac. The moon and planets also orbit within this plane, so they too move along the zodiac.

Figure 4 shows the orbit of the Earth and the apparent path of the sun relative to the background stars. An arrow shows the position of the sun for January/February. The sun is in the constellation of Capricornus. These stars are in our daytime sky and cannot be seen. Opposite the sun, and visible throughout the night, is Cancer. Six months later, in July/August, the sun will be in Cancer, and Capricornus will be visible in our night sky.

It is this, the location of the sun in the zodiac at the time of your birth, which determines your 'star sign' in astrology. Each year for the Phoenix Astronomical

4 The apparent paths of the sun and moon along the zodiac.

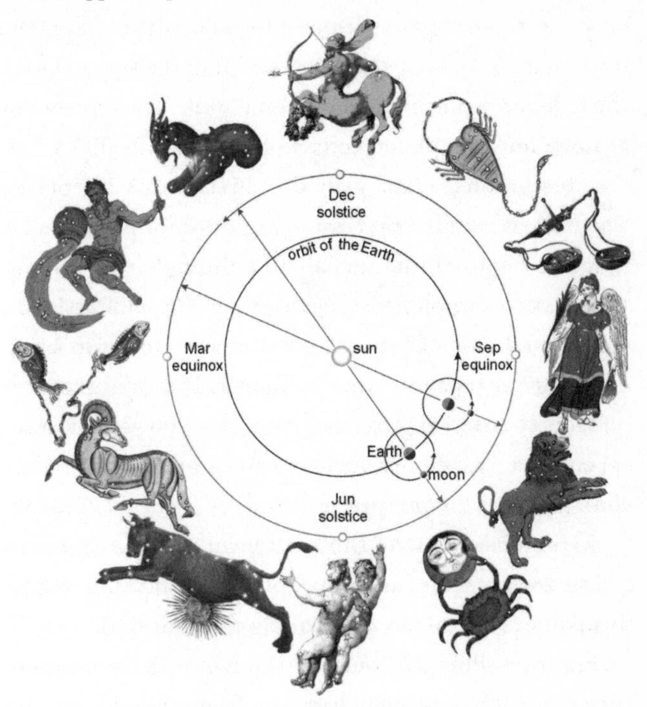

Society, I and others produce the *New Zealand Almanac* which, in calendar form, provides information on the sun, moon, stars and planets for the year. This includes the sign of the zodiac in which the sun is situated from month to month. Each year we get calls from readers who tell us we've got it wrong because the dates don't correspond with the dates given for their star sign in magazines and newspapers.

How can this be? Well, the Earth is slowly wobbling on its axis like a giant spinning top. The tilt remains the same at 23.5 degrees, but the position in the sky to which the poles are pointing is slowly changing. The movement is slow: one complete cyclic wobble takes 26,000 years.

This gradual change in the direction of tilt of the Earth's axis is called precession. An effect of precession is that the signs of the zodiac slip through the seasons. Today, for example, the northern spring equinox is in the constellation of Pisces; 2,000 years ago it was in Aries, and 5,000 years ago it was in Taurus. It moves along by one constellation approximately every 2,200 years. Consequently, the meanings given to the signs of the zodiac changed with the passage of time.

Most modern astrologers know very little astronomy, but in ancient times astrologers updated their star charts to take account of precession. During the dark ages (a term invented in the fourteenth century to describe the preceding 900 years in Europe, from the fall of the Roman empire to the beginning of the Renaissance) this practice ceased. As a result, astrological charts and dates used by most astrologers today are more than a thousand years out of date. According to astrological charts in my local newspaper I am a Scorpio, but in fact the sun was in the constellation of Libra when I was born.

Those familiar with the musical *Hair* will have heard of 'The dawning of the Age of Aquarius', but what does it mean? I heard an astrologer on the radio who, when asked when the Age of Aquarius would begin, answered, 'I think it is happening now.' It's not. The Age of Aquarius will begin when the vernal equinox, the northern spring equinox, moves into Aquarius and that won't be for another 600 years.

The most

incomprehensible

thing about the

universe is

that it is

comprehensible.

Albert Einstein

The first computers

MANY YEARS AGO I went with friends on a camping holiday in Cornwall, England. We drove through the night, and just before dawn as we were travelling over the Salisbury Plain we saw a sign, 'Stonehenge'. It was still dark when we got out of the car and walked along the path. As we approached I saw the huge stone giants silhouetted against the faint glow of early twilight. This was before Stonehenge was fenced and there, in the still of dawn, I could feel the past: it was as if the stones touched me.

For millennia people have gazed in awe at Stonehenge, often totally unaware of how such structures were used. Astronomy is the oldest of sciences, and knowledge of the sun, moon and stars' daily and seasonal changes was essential to the survival of early communities. Structures such as Stonehenge – and other mysterious edifices found around the globe – revealed the predictable nature of the cosmos to our ancestors. In this sense they were the first computers.

There is no written record of the people who built Stonehenge four to five thousand years ago – they are simply known as the 'Beaker People', because of their strange tradition of burying pottery drinking cups with their dead – so we will never know for certain all of the purposes for which the ring of giant stones was used. However, the stones have so many alignments associated with the cycle of the sun and moon that these are unlikely to have occurred by accident. Using the stones these early Britons would, among other things, have been able to predict the occurrences of solstices and equinoxes.

The Earth's axis is tilted 23.5 degrees to the plane of the solar system, and remains fixed and aligned with the celestial poles as the Earth orbits around the sun. It is this tilt that gives us our seasons. Figure 5 shows the Earth at four points in its orbit around the sun. In December the Southern Hemisphere is tilted towards the sun: it is mid-summer in the Southern Hemisphere, mid-winter in the Northern Hemisphere. Six months later, in June, the situation is reversed and the Southern Hemisphere is tilted away from the sun: it is mid-winter in the Southern

5 The Earth's tilt and the seasons. As the Earth orbits around the sun, the north and south poles are alternately tilted towards the sun. The sun's altitude therefore increases and decreases during the year, producing seasons.

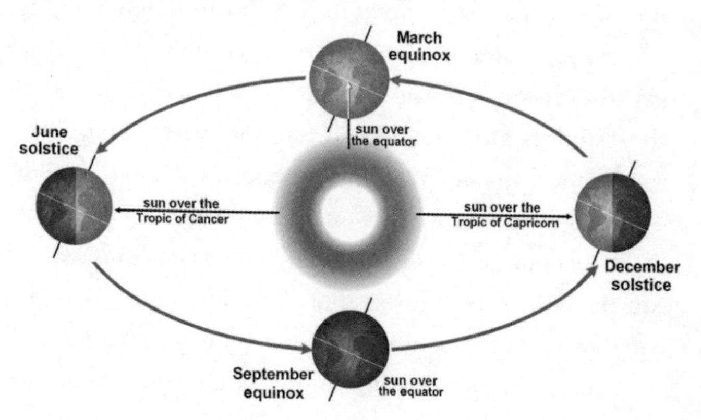

Hemisphere, mid-summer in the Northern Hemisphere. These two points, which mark mid-winter and mid-summer, are called the solstices.

There are two midway points, one in March and another in September, when neither hemisphere is tilted towards the sun. These are the equinoxes. March is the Southern Hemisphere autumn and Northern Hemisphere spring. September is the Southern Hemisphere spring and Northern Hemisphere autumn.

Due to the tilt of the Earth, the path of the sun – the ecliptic – is tilted 23.5 degrees from the celestial equator, and intersects it at two points, the equinoxes (see figure 2, page 47). Over the course of a year, the sun will appear to move from one side of the celestial equator to the other. Consequently, throughout the year the rising and setting positions of the sun, its altitude at noon, and the number of hours of daylight are continually changing.

At our winter solstice (figure 6, A) the sun rises at its greatest distance north of east. Its path across the sky is short and its altitude at noon is at its lowest. This is the shortest day and in Wellington where I live, at latitude 41 degrees south, we get only nine hours of daylight.

At an equinox (figure 6, B), and only at an equinox, the sun rises due east and sets due west. The hours of night and day are equal, so we get twelve hours of daylight.

At our summer solstice (figure 6, C) the sun rises at its greatest distance south of east. Its path across the sky is long and its altitude at noon is at its highest. This is the longest day and in Wellington we get fifteen hours of daylight.

By marking the positions at which the sun rose or set along the horizon, people could determine the exact dates of the solstices and equinoxes, and so predict the changing of the seasons. Stonehenge is, then, an elaborate structure built to mark these positions.

My moving encounter with Stonehenge in England led me, many years later, to conceive the idea of creating the same structure in New Zealand, adapted to Southern Hemisphere conditions. As I write this book, Stonehenge Aotearoa is nearing completion. On a flat hilltop in rural Wairarapa, a group of passionate astronomers has created

6 Path and altitude of the sun at latitude 45 degrees south at: (A) winter solstice, (B) an equinox, and (C) summer solstice.

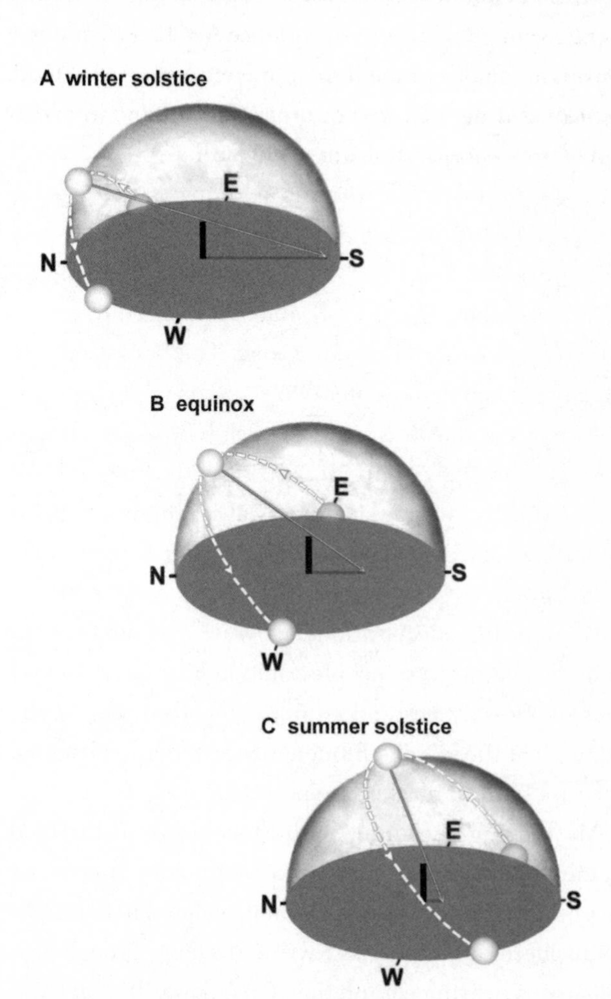

a working henge, built on the same scale as the English structure before it fell into ruins. It is intended to inspire people, young and old, to experience for themselves how the remarkable technologies of ancient times were used, and can still be used, to get practical, detailed information on the seasons, time and navigation.

Things that go bump in the night

TWENTY THOUSAND years ago, bands of hunter-gatherers would meet to exchange goods (and probably people). Although there were no settled communities, the first trade routes had been established. A meeting place could be arranged, but without clocks or a calendar how did they appoint a meeting time? It so happened that there was, and is, a clock in the sky. Wolves howl at it, lovers meet under it and poets write about it. We call this heavenly lantern the moon.

The moon goes through a regular, clock-like cycle but these days many people find its movements mysterious.

Why, for instance, do we see the moon in the morning some days and not on others? Why, when we can see half of the moon, do we call it a quarter moon? These are tantalising questions.

As the moon orbits the Earth we see differing amounts of its dark and illuminated sides. These are known as phases of the moon (see figure 7). The moon's cycle, from one 'new moon' to the next, takes 29.53 days, or approximately one month using our present calendar. This is where the word month – a moonth – comes from.

At the beginning of each cycle we first see the 'new moon' in the west, shortly after sundown. It is a thin crescent, with its 'dark side' faintly illuminated. The bright crescent is illuminated by sunlight. The dark side is illuminated by reflected light from the Earth. If you were standing on the dark side of the moon at this time, you would see the Earth almost full and it would, through light reflected from the sun, be illuminating the lunar landscape.

Night by night the moon moves eastward away from the sun, and sets later and later. As it does so it waxes: we see more and more of the sunlit side of the moon. The moon's crescent points westward, which is the direction of the sun. Eventually we see half of the moon illuminated – which we call a 'first quarter moon' because the moon has completed a quarter of its cycle. At first quarter the moon rises close to midday, is due north at sundown, and sets at midnight.

7 The phases of the moon. The large images of the moon show the phase visible from Earth at different points in the moon's orbit.

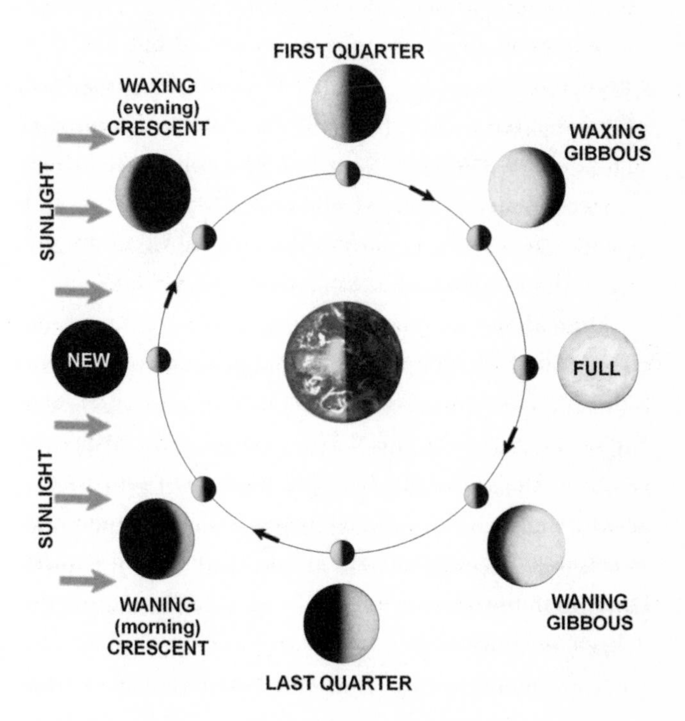

The moon continues to move eastward and we see a gibbous moon, one that is between half and fully illuminated. The word gibbous comes from the Latin word for hump; the illuminated part of the moon appears as a protuberance, bigger than a semicircle and less than a circle. Eventually we see a 'full moon'. The full moon is directly opposite the sun in the sky. It rises as the sun sets. It is then visible all night, and it sets at dawn. The moon has now completed half of its cycle.

The moon continues to move eastward but it is now waning, moving back towards the sun. Eventually it becomes a half moon again, but this time its illuminated side points to the east. As it has now completed three-quarters of its cycle it is called, at this stage, the 'last quarter moon'. At last quarter the moon rises at midnight, crosses the meridian at dawn, and sets at midday.

Thereafter the moon continues to wane to a thin crescent that is eventually invisible in the morning twilight. It is lost from view because it is close to, and rising and setting with, the sun. Three nights later it reappears in the western evening twilight: we have a new moon. Scientifically speaking, a new moon is when the moon is in conjunction with the sun – that is, it is between the Earth and the sun – and can't be seen. Traditionally, though, we call it a new moon (and bow three times and turn our money over) when we first see it as a thin crescent at the beginning of its cycle.

The moon's regular cycle is so easy to observe, it is not surprising that people used it as the first clock. Two nomadic groups could arrange to meet at a certain place in, say, four months' time. All they needed to do was count the full moons. If it would take two weeks to travel to the meeting place, they would embark on the journey at the time of the new moon before the fourth full moon.

The first calendars were based on the cycle of the moon. They divided the year into twelve months, with each month divided into the four quarters, or 'weeks', of the lunar cycle. It seems likely that the reason we have twelve constellations in the zodiac is that there are twelve full moons in a year.

When the moon is full, it is directly opposite the sun (see figure 4, page 55). Each month, as the sun moves from one constellation of the zodiac to the next, so does the location of the full moon. Even though you cannot see the stars in the daytime, you can always tell which constellation the sun is in by looking at the full moon: it will be in the zodiacal constellation directly opposite the location of the full moon. For example, if the full moon is in Pisces the sun will be in Virgo.

Today most western societies use a solar calendar. This is not a new idea: the Babylonians had a solar calendar from around the seventh century BC, and our version was introduced by Julius Caesar in 45 BC. The Roman emperor even re-named a month after himself (and stole an extra day from February to make it longer), but lived only a year to enjoy the honour before being assassinated. Augustus, who followed Julius, also had a month named after him. He couldn't have one with fewer days than Julius, so he also took a day from February. Thus our present calendar is the result of tinkering by various Roman emperors.

In the solar calendar the year is still divided into

twelve months, but these are no longer in step with the cycle of the moon. Usually we have twelve full moons in a year, although every two to three years we get thirteen. When this happens, one month will have two full moons instead of one. The second is called a 'blue moon', hence the expression 'once in a blue moon'. Note, however, that lunar calendars are still used today by Jewish and Islamic cultures. Matariki, the Maori New Year, is also based on the lunar calendar.

There is something magical about the moon that touches the human soul. In ancient Mesopotamia it was closely associated with the great Earth-mother goddess and her cycle of life, death and rebirth. At the beginning of each cycle the moon is born in the glow of evening twilight. It then grows to full maturity, before declining and finally being lost in the early light of dawn. Many ancient peoples thought it had been consumed by the eternal fires of the sun. When it reappeared they believed it was indeed a new moon. Old habits die hard: in the twenty-first century we still talk of a new moon, even though we now know that it's the same old moon.

The moon has long been associated with ghosts, witches and things that go bump in the night. Traditionally, the supernatural powers of the underworld were believed to wax and wane in concert with the moon. Even today we always have a full moon in a good gothic horror movie, as a sure sign werewolves and vampires are abroad.

Like the moon's phases, the great Earth-mother goddess associated with the moon has come in many forms. To the ancient Greeks she was Hecate, the goddess of the underworld, who roamed the night accompanied by her legions of ghosts and vampires. Her name was spoken with a hushed voice and then only by women.

It was believed that part of this great goddess was within every woman. The moon's cycle was seen as a symbol of the female cycle, and of women's magical power to give forth life.

Hecate was also the goddess of witchcraft: if called upon, she would bestow powers on women. Not surprisingly, men feared her. The Romans believed she wore a necklace made from men's testicles around her neck. In Europe as recently as 150 years ago, offerings were left for her at crossroads, the places where people's paths would meet, and where executions were carried out.

You will have noticed that the moon has light and dark markings (figure 8, A, following page). The dark regions are called mare, the Latin word for sea, because in the early days of the modern era the moon was believed to be similar to the Earth. Space probes starting from the 1960s have shown there is, in fact, no liquid water on the moon. However the light and dark markings do bear a similarity to the Earth's geology. The mare are vast basins and plains of basaltic lava, while the lighter areas are highlands. The Earth also has huge basaltic basins, but these are filled with water and form the great oceans.

8 The light and dark markings on the moon: (A) Southern Hemisphere view, (B) Northern Hemisphere mirror image view.

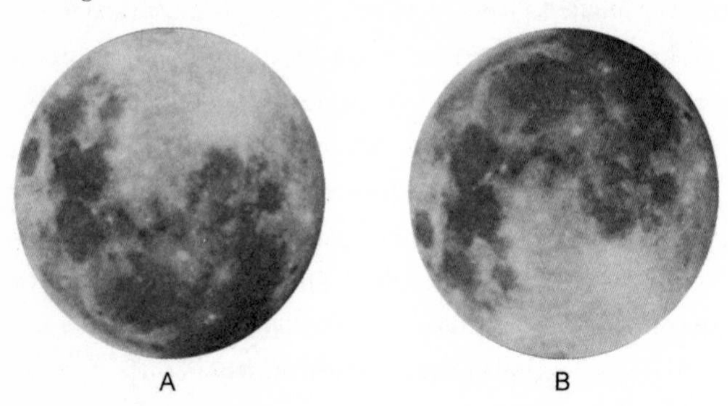

The pattern of light and dark markings form the famous 'man on the moon', which I personally have never been able to distinguish. When trying to look for this figure, remember that the moon appears upside down to how it is seen in the Northern Hemisphere. To Maori the markings represent Rona who, because she swore at the moon, was transported to its surface, along with the gourd she was carrying and the punga tree she was holding on to. However, after long observation, what I see on the moon is a rabbit. See if you can pick it out.

One of the largest and darkest regions of the moon is the Mare Imbrium. Ancient Greeks knew this as the Plains of Persephone and believed it was here, at the shrine of Hecate, that the souls of the dead were judged. By comparison, no living soul could see the Elysian

Fields where the virtuous would find peace and happiness after death: they were on the far side of the moon, facing heaven.

According to the medieval Christian church, both the sun and the moon were perfect and unblemished. One of the reasons Galileo got into trouble was because he reported seeing spots on the sun. But how could anybody think the moon was unblemished when its mottled appearance was plain to see? The explanation was that the moon was a mirror, and the dark markings merely reflections of the Earth's imperfections.

To see the Earth on the moon, you have to look at a mirror image of the moon as seen in the Northern Hemisphere (see figure 8, B). You will detect a certain likeness to the then known world – for example, the shapes of Spain, the Mediterranean, North Africa and India. Hard though it is to imagine now, people looked at the moon in a mirror to try and map regions of the world that had not yet been explored. If you look at the part of the moon that is supposed to represent Africa, you will see that its southern extent is divided by sea. This sea, which of course does not exist, is shown on some atlases drawn up in the thirteenth century. Note also that in what is supposed to be the Atlantic Ocean there is a large circular land mass, Mare Crisium. Was it thought that this was the fabled lost continent of Atlantis?

A simple pair of binoculars reveals a wealth of detail on the moon; in a large telescope it is stunning. A description of the moon's magnificent geography is beyond the scope of this book but I can tell you I never tire of looking at the awesome surface features of this alien world, and floating silently above its dark plains, with their craters, cracks, and crevasses, and onward, over its rugged, cratered highlands.

The wandering stars

IF, ALONG THE ZODIAC, you see a bright star not marked on your star charts, it will almost certainly be a planet. Five planets are easily seen with the naked eye, and some outshine the brightest stars. All the planets revolve around the sun in the same direction – counter-clockwise when viewed from a point north of the solar system. From the surface of the Earth they are seen to slowly move against the background stars. Indeed, the word 'planet' means wandering star.

Despite their name, however, planets are not stars, but neighbouring worlds that shine only by reflecting the light of the sun. Because their distance from the Earth is continually changing, so is their brightness. After observing for a while, you will be able to identify each planet by its individual brightness and hue.

How a planet appears to move in the sky depends on whether it is inwards or outwards of the Earth. Seen from Earth the inner planets, Mercury and Venus, appear to spend most of their time close to the sun. When one of them is at its closest point to the Earth (C in figure 9) it is invisible. Not only is it close to the sun but, like the new moon, its dark hemisphere is turned towards us, a position known as 'inferior conjunction'. Conjunction describes the situation where two celestial objects, usually the sun and a planet, are close together in the sky.

The closer a planet is to the sun, the greater its speed and the shorter its path. At inferior conjunction, it will overtake the Earth in the inside lane. As it moves towards point W it will rise in the east, just before the sun. Each morning it will then appear to move further westwards, rising a little earlier each day, until it reaches what is called 'western elongation'.

The planet will then appear to reverse direction and move back towards the sun. Soon it will be lost in the twilight as its journey takes it *behind* the sun – a position called 'superior conjunction' (D). Next the planet will appear on the other side of the sun, in the western evening twilight. It will set later and later each evening until it reaches 'eastern elongation' (E). Once again it will reverse direction and appear to move back towards

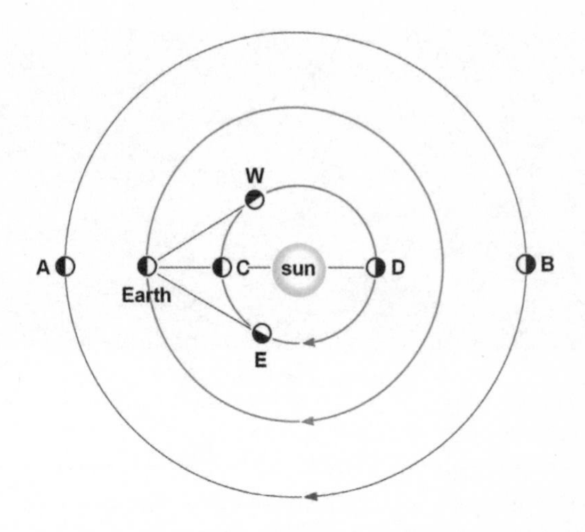

9 Planetary phenomena: the orbital positions of an outer planet at (A) opposition and (B) conjunction; and for an inner planet at (C) inferior conjunction, (D) superior conjunction, (E) eastern elongation, and (W) western elongation.

the sun, until it is again lost in its rays. Mercury and Venus are best seen at or near elongation – their greatest apparent distance from the sun.

Note also that the inner planets, just like the moon, exhibit a cycle of phases as we see differing proportions of their illuminated side (figure 10, following page). Mercury's phase cannot be seen without a telescope, but Venus's can, on occasions, be seen with binoculars. The best times to look are before and after inferior conjunction, when Venus is close to the Earth and therefore larger. Observe it in a bright twilight sky when its glare

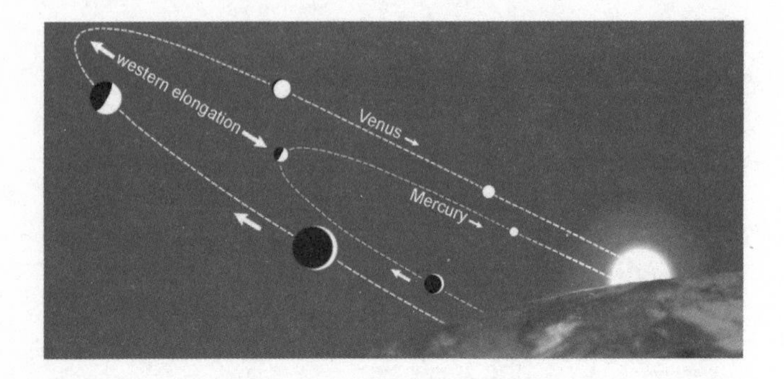

10 The apparent paths and phases of the inner planets. For a morning (eastern) appearance, the planets appear from in front of the sun and move to its left, and their movement is away from us. For an evening (western) appearance the sequence is reversed: the planets appear from behind the sun and move to its right, and their movement is towards us.

is subdued. **NEVER LOOK FOR A PLANET WITH BINOCULARS IF THE SUN IS IN THE SKY**: if you accidentally look at the sun, you will suffer serious eye damage.

Mercury, associated with the fleet-footed winged messenger of the gods (known to the Greeks as Hermes) is a small world that orbits around the sun in just 88 days and makes fleeting appearances in the morning and evening twilights. The best time to see this planet from the Southern Hemisphere is when the sun is close to an equinox. This occurs in March, when Mercury can be found in the morning sky, and September, when it is in the evening sky. Mercury has a slight pinkish hue, and in twilight its bright lustre is subdued.

Venus, on the other hand, is the brightest star-like object in the sky, outshone only by the sun and moon. Traditionally associated with the goddess of love and sex, known variously as Aphrodite (Greece), Ishtar (Babylon), Astarte (Syria) and Venus (Rome), it is brilliant white at maximum brightness, and can, if you know where to look, be seen in broad daylight. In any one year Venus is either in the early morning or early evening sky, hence the popular names Morning Star and Evening Star. In ancient times the two manifestations were thought to be different planets, which the Romans called Hesperus and Phosphorus.

At its maximum elongation Venus is 48 degrees from the sun, and can be seen three hours before or after the sun is in the sky. For early Polynesian sailors, it was an important navigational beacon: because of its brilliance it could be seen in the twilight, when no other stars were visible.

The outer planets – of which the brightest are Mars, Jupiter and Saturn – can be seen at any time of the night. Their slow orbital motion, like that of the moon, is in an eastward direction. Each appears brightest when closest to us – that is, when the Earth lies between the planet and the sun, a point called 'opposition' (A in figure 9). At opposition the planet is opposite the sun, and will rise as the sun sets.

As an outer planet approaches opposition, it undergoes a curious change in its path against the background stars. Its eastward motion appears to slow, halt, and then change to a westward motion. After opposition the westward motion stops, and the planet reverts to its normal easterly motion. This is called 'retrograde motion'.

Retrograde motion puzzled ancient astronomers, who believed the Earth to be stationary at the centre of the universe. The explanation is that the Earth has a smaller orbit and higher orbital velocity, and therefore overtakes a superior planet. Before this the planet appears to be moving eastward, in the same direction as the Earth. As it is passed by the Earth, however, it appears to move backwards relative to the background stars. After opposition it appears to be again moving in the same direction as the Earth.

You can imagine this effect as being similar to what you see when you overtake another car. At first, as you approach the slower car ahead, it appears to be moving in the same direction as you, relative to the background scenery. As you overtake it, it appears to move backward relative to the background topography. After you have passed it, although falling behind it will once again appear to be moving in the same direction as you, relative to the background.

Mars, named for the Roman god of war, varies greatly in brightness depending on its distance from the Earth. It is prominent in our sky once every two years, when it looks like a bright orange-red star. By comparison Jupiter, which looks like a brilliant yellow-white star, glows brightly every year. As a celestial beacon it is rivalled only by the sun, moon, Venus, and occasionally Mars. It moves sedately among the stars, taking 11.9 years to complete one orbit of the sun.

Jupiter was very significant in ancient astrology as it was associated with the kings of both gods and men. It has more than 60 moons, of which four – Io, Europa, Ganymede and Callisto – can be seen with binoculars. The best time to look is in twilight. The moons will look like tiny stars in a line, and from night to night they will change their positions as they orbit around the planet.

Saturn was the most distant planet known before the invention of the telescope. It moves very slowly, taking 29 years to complete one revolution around the sun. In ancient times Saturn was associated with Chronos, the god of time, because the time it took to revolve around the zodiac equalled the average life expectancy of human beings. If, for example, you were born when Saturn was in Capricorn, you were on borrowed time if you survived after its return to this constellation. Today most New Zealanders can expect Saturn to complete two laps of the zodiac during their lifetime, and a few observe three. It takes Saturn more than two years to move from one zodiacal constellation to the next. The faintest of the naked-eye planets, it looks like a bright yellowish star. With its magnificent rings it is one of the most spectacular of telescopic objects.

Beyond Saturn are three more planets. Uranus can be seen as a faint star from a dark sky site, if you know where to look. Neptune can be seen with binoculars but you will need a chart to find it. Finally there is the tiny frozen world of Pluto, which can't be seen in anything less than a fairly large telescope.

Beneath the Milky Way

WHEN YOU FIRST look at the night sky, particularly on a dark moonless night in the countryside, there seem to be so many stars it may appear an impossible task to fathom it all out. But it's really no different from learning the geography of the world, or even finding your way about your home town when you were young. Most of the objects in the night sky have a fixed location, so if you start off by identifying the continents – the bright stars and planets – the rest will fall into place with time and practice. After a while you will be able to identify a particular star or planet simply by its tint and lustre.

The best time to start learning to read the night sky is in the evening twilight, when the bright stars are visible but the sky is not crowded with fainter stars. The first thing that's apparent is that the stars differ in brightness. This may seem obvious but it appears to have been missed by many movie-makers, who often create an artificial night sky – or starry background to a spacecraft – with stars all of the same brightness.

In 129 BC the Greek astronomer Hipparchus divided the stars into six orders of brightness, which he called magnitudes. The brightest stars were placed in the first magnitude. The faintest that could be seen with the unaided eye were placed in the sixth. A modified version of this magnitude scale is still used in astronomy today.

Not only do the stars differ in brightness but their numbers increase with increasing faintness. For example, there are far more stars of the third magnitude than there are of the second. On a clear, moonless night, well away from city lights, you may see as many as three thousand stars with the naked eye. A look through binoculars will reveal tens of thousands.

Now look at the brightest stars. Sometimes they will appear to flicker or twinkle. This twinkling is due to the refraction, or bending, of the stars' light as it passes through the turbulent atmosphere. Seen from space, or directly overhead on a calm night, the stars are still and motionless. When a star has just risen, or is about to set, and is close to the horizon it may flash different colours.

When we look towards the horizon we are looking through the thickest and densest part of the atmosphere and the white starlight is being broken into a spectrum of colours, like a tiny rainbow. Later, when the star has risen higher in the sky, the twinkling will stop or become less pronounced.

Looking at the brighter stars you will notice that they have different colours. These colours are subtle tints, rather than bold colours like traffic lights. The colour of a star, which can help identify it, is directly related to its surface temperature. Our sun, around which our planet Earth orbits, is a typical star: it appears much bigger and brighter than the other stars only because it is comparatively closer to us. The sun is a yellow star. Stars that are hotter than the sun look white, or bluish. Cooler stars look orange or red.

On the subject of colour it is useful to know that the human eye uses two different types of light-receptive cells: rods and cones. The cones give us colour vision but are not as sensitive as the rods, which provide us with black and white vision. At night, when light levels are low, we are generally limited to seeing in black and white. Even the normally brilliant colours of flowers are reduced to greys. This is why night scenes in a colour movie don't look right: a camera can see colour but your eyes can't. It is also why spooky movies are more effective in black and white: the brain knows that it is more like the real thing.

When you look at the stars at night most of them look white, because there isn't enough light entering your eyes to trigger the cone cells. If you have the opportunity, look at the Milky Way at low power through a large telescope. You will discover that the stars look like glowing gems – amethysts, diamonds, sapphires and rubies. Because the telescope gathers more light, you're suddenly seeing in colour.

Our sun is a single star, but about half of all the stars in the sky are multiple systems. Binary systems – two stars orbiting around each other – are very common. Triple, quadruple and even sextuplet systems can also be found. The stars of a multiple system are so close together that to the unaided eye they look like a single star. However, a good astronomical telescope will resolve thousands of apparently single stars into doubles and triples. If the stars in a binary system have very different temperatures, the colours can be stunning. For example, the star Albireo in the constellation of Cygnus actually consists of a glorious golden-yellow star with a sapphire-blue companion. Through a telescope the pair look like celestial traffic lights.

Let's return to gazing at the stars with the unaided eye. What's immediately obvious is that the bright stars are not distributed randomly across the sky. Most of the brighter ones congregate together to form a band that stretches across the sky. This band marks the path of the Milky Way. Like a huge celestial vertebra it forms the backbone of night. In a twilight sky, or from a city where the Milky Way is not usually visible because of bright lights on the ground, you can always work out where the Milky Way is by identifying this path of bright stars. On a clear dark night the Milky Way can be seen as a gossamer of faint ghostly light arching across the heavens, its path sown with stars. Its orientation varies from season to season; during early evening in summer it runs along the horizon.

Since earliest times people have wondered: What is this path of light? To the ancient Egyptians the Earth was a mirror of the heavens, and the Milky Way the celestial equivalent of the River Nile. The ancient Greeks had a different theory. They believed that Hera, Queen of Gods, had been tricked into feeding the infant Hercules, child of her husband's mistress. When she discovered the infant's true identity, she cast him down and her breast milk spilled across the heavens, forming the Milky Way. To the native North Americans the Milky Way was a celestial highway along which the dead travelled on their way to the happy hunting grounds; the stars were the travellers' campfires.

In the Southern Hemisphere we have our own unique tale. To early Maori the stars were eyes without bodies. These eyes originally lived in great darkness at a place called Mount Maunganui. Then Tamarereti placed them in his canoe, Uruao, and carried them up to the sky. After placing the important navigational and seasonal stars in their places, he tipped the canoe upside down and the remainder of the stars spilled out across the sky to form Te Ikaroa, the Milky Way.

The true nature of the Milky Way can be uncovered with a simple pair of binoculars, which will resolve the faint light into thousands of stars. Our sun, along with all of the other stars we see in the night sky, is part of a gigantic system of stars called a galaxy. This galaxy is a huge, flattened, spiral system, one hundred thousand light-years across, and containing more than one hundred billion stars. That's worth a thought if you're watching *Star Trek*. Even if you could travel at the speed of light, it would still take you 100,000 years to travel from one side of the galaxy to the other.

The Milky Way is, essentially, the galaxy seen edge-on. Its light comes from millions of distant stars – too far away and too faint to be seen individually. The stars we can see with the naked eye are our neighbours.

Signposts in the sky

LET'S START TO explore the night sky. It's a good idea to start on a summer evening. First, it's relatively warm then compared with the rest of the year. And, secondly, at this time there is a great celestial signpost in the sky. And a signpost is just the thing you need if you are unfamiliar with the night sky.

Figure 11 on the next page shows the constellation of Orion, The Hunter, easily recognisable by its line of three bright stars. These three stars, known to Maori as Tautoru, form the Belt of Orion. They also form the bottom of the Pot. Arrows on the chart show how the

11 The Orion signpost.

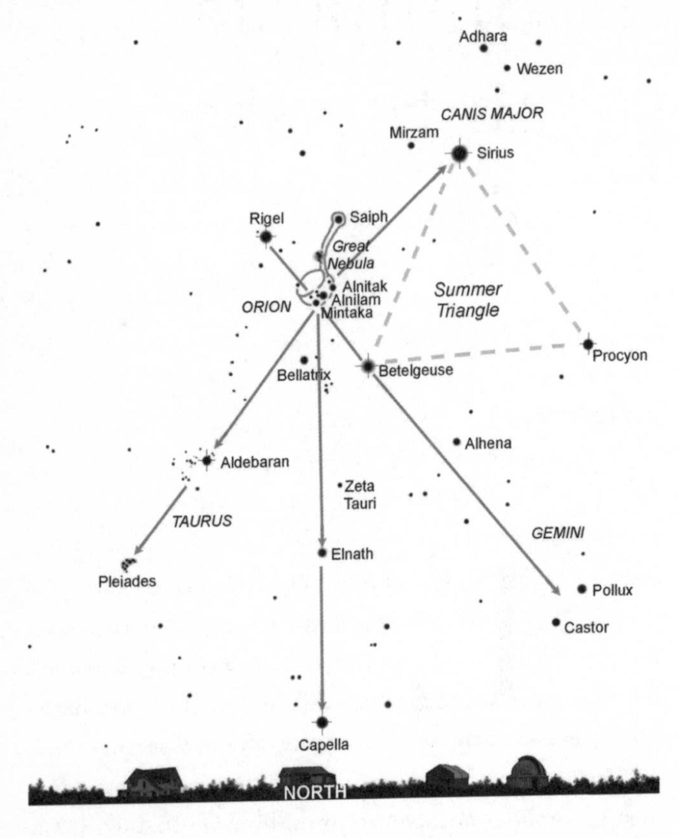

bright stars of Orion may be used as a signpost to identify other bright stars in the sky. Orion can be found in the morning sky from mid-June to December, and in the evening sky from January to mid-May. At the beginning of the year it is due north at midnight. If you take a look at the star charts at the end of the book, you will notice that Orion, along with other constellations originally named in the Northern Hemisphere, appears upside down in our southern night sky.

When we look at bright stars like Orion, we are not seeing the common or garden variety of star in the Milky Way. Most of the bright stars are giants – hundreds, sometimes thousands, of times brighter than the sun. (Our sun is used as a standard candle against which we measure the brightness of other stars.) These giant stars are both remote and rare, yet they dominate our night sky. Like great cosmic lighthouses, they blaze out across the light-years.

The brightest star in Orion is the steely blue-white star Rigel. Rigel is 770 light-years away but ranks as the sixth brightest star in the sky because it shines with the light of 40,000 suns. If Rigel were the sun of *our* solar system, all of the planets – including Pluto, the most distant – would vaporise. Rigel is, in fact, the brightest member of a complex system of five suns. It is twice as hot as our sun and has a diameter 66 times larger.

If you think that's large, consider the second brightest star in Orion – Betelgeuse, the Shoulder of the Giant. You might remember a movie called *Beetlejuice*, in which the lead character was a demon by the same name. If you look hard at Betelgeuse (Beetlejuice), pronounced Bet-el-gurz, you will notice that it has an orange-red tint. In ancient times red stars, not being pure and white like other stars, were associated with devils and demons. Like the planet Mars, believed to be the bringer of war and destruction, they were tainted with a bloody hue. Even more ominously, Betelgeuse, like Mars, waxed and waned in brightness.

Mars, a small nearby planet, changes in brightness according to how far it is from the Earth at any given time. Betelgeuse, on the other hand, is a colossal star that fluctuates in brightness because it is pulsating. Although, area for area, its surface doesn't radiate as much energy as the sun's, its size more than makes up for it. Betelgeuse is so large that if you placed it where the sun is it would engulf all of the planets out to and including Mars. It belongs to a class of star known as 'red supergiants', and is so huge it is unstable. Its instability causes it to pulsate in size to diameters between 300 and 400 times that of the sun. As it does so, it fluctuates to between 4,300 and 14,300 times the brightness of the sun.

At this stage you might be thinking that our sun is, on a cosmic scale, a small feeble star. Fortunately for us, this is not so. It is, in fact, above-average in size and brightness. Of the 27 stars within 12 light-years of the sun, only three are brighter and only nine are bright enough to be seen without a telescope. High luminosity stars are rare. Only one star in a million is as big as Betelgeuse, or as bright as Rigel.

The star Bellatrix marks the other shoulder of the giant. Bellatrix is called the Amazon Star because, according to stories which have come down from antiquity, women born when it was rising would be great warriors and leaders. Bellatrix is a blue-giant star 360 light-years away and 3,000 times brighter than the sun.

At a distance of 1,300 light-years, Saiph – the name is Arabic for sword (most star names are Arabic because Arabs produced the best, most accurate early star charts) – is the most remote of the bright stars in Orion. Known as the Eye of the Tiger, it is another blue-giant star, 20,000 times brighter than the sun.

The three bright stars that make up Orion's belt, Mintaka, Alnilam and Alnitak, are ferociously hot giant stars. In fact, if you were to place the sun alongside any one of these stars, it would appear almost completely black. These stars are so hot that most of the energy they emit is ultra-violet radiation.

It's a strange

courage

you give me

ancient star:

shine alone in the

sunrise

toward which you

lend no part!

William Carlos Williams, 'El Hombre'

Forge of the gods

BETWEEN THE STARS and along the tracks of the Milky Way swirl cosmic dust and gas. In many regions this dust and gas collects into vast clouds, which can be seen as dark patches silhouetted against the bright background of the Milky Way. This mosaic of starlight and dark clouds gives the Milky Way its mottled appearance. These dark interstellar clouds are the raw material from which new stars and planets are born. It is estimated that about six new stars are born in the Milky Way galaxy each year.

According to ancient Greek mythology, at the beginning of time, before the gods, the first beings to come into existence were the Titans. The Titans were the personifications of the uncontrollable forces of nature. Among their creations were the Cyclops, three huge giants, each sporting at the centre of their forehead a single eye that shone like a brilliant star. The Cyclops worked at the forge of the gods, fashioning the raw material of the universe to give it form and beauty.

The secrets of this forge, the actual processes of star birth, are normally hidden from view, buried deep within the dark clouds. But there is a place where the Cyclops have left the door open, where we can look inside and see the Titans' secrets. Fifteen hundred light-years away, just beyond the bright stars that form the constellation of Orion, there is a gigantic dark cosmic cloud. The presence of this dark cloud is the reason the Milky Way is not as bright near Orion and Taurus as it is everywhere else.

Less than a million years ago a gravitational collapse within the cloud led to the explosive birth of a number of super-massive stars. Intense radiation from these stars formed an expanding bubble of hot gas, and this ruptured the wall of the cloud. With the unaided eye, on a clear, dark, moonless night you will see a faint glow coming from the Sword of Orion – centred on the central star that forms the handle of the Pot. This is the Great Nebula in Orion, a glowing crater in the wall of the dark cloud. Seen through a large telescope it is one of the most breathtaking sights imaginable, a churning, turbulent star factory set within a maelstrom of flowing, luminescent gas. Even binoculars will reveal some of the wreaths and swirls of glowing gas. Within this forge of gods the gas is heated to a temperature of 10,000 degrees Celsius by a torrent of energetic ultraviolet light from the four hottest and most massive stars in the Orion nebula. These stars, known as the Trapezium, are visible in small telescopes.

As well as the four bright Trapezium stars (two of which are in fact binaries, or double stars), three other stars belong to the central cluster. Hence the Trapezium cluster is actually nine stars. These nine stars have a combined mass of about 110 suns, but occupy a volume of space only 0.1 light-years across. That's one forty-third the distance between the sun and the star nearest to it.

The closeness of the Trapezium stars shows that they are bound together gravitationally, and are moving in complex orbits around each other. As these orbits are constantly changing, the entire Trapezium system is highly unstable and it is only a matter of time before close encounters occur. Whiplash can then hurl stars out of the system.

There is strong evidence that the Trapezium cluster has been in a continuous state of disintegration since its birth. Eleven other large stars have been identified as moving directly away from the Orion nebula; it appears that they have been hurled out of the nest. One of these young runaway stars, AE Aurigae, has already travelled

44 degrees across the sky from its birthplace. Its path has taken it almost directly north out of Orion, through the constellation of Taurus and into the constellation of Auriga. It appears to have been ejected from the Orion nebula less than a million years ago.

As the hot bubble of gas that formed the Orion nebula expanded, it compressed the walls of the cloud, generating a second wave of star birth. In addition to the stars of the Trapezium, the cavern contains 700 other young stars in various stages of formation. With time, a cluster of thousands of stars will emerge from what is now the Orion nebula. Unlike the giants of the Trapezium, these second-generation stars appear to be similar in mass to our sun. A disk of material in which planets appear to be forming surrounds each of these stars. New worlds are in the making.

The Seven Sisters

ALL STARS ARE formed in clusters, but with the passage of time the clusters disperse; few survive more than 1,000 million years. By comparison, our sun and its system of planets is 4,600 million years old. Once, long ago, our sun was part of a star cluster, but its siblings have long since vanished and now make their own way through the star-fields of the galaxy. If you look along the Milky Way with a pair of binoculars, you will discover dozens of star clusters. By their very existence you will know they are composed of comparatively young stars.

The brightest and most famous star cluster in the sky

is the Pleiades, known in folklore as the Seven Sisters, and to Maori as Matariki. The pre-dawn rising of Matariki occurs in early June, and heralds the Maori New Year. To find the Pleiades use the 'Orion signpost' on page 90.

Take a path left along the line of the three Belt stars. First you will encounter the bright orange star Aldebaran, the 'follower' of the Pleiades. Tarry a while here. Surrounding Aldebaran is a cluster of about 200 stars, the brightest of which are visible to the naked eye and form a distinctive V pattern. These are the Hyades. In Greek mythology they are the daughters of the Titan Atlas, and so are half-sisters of the Pleiades. Atlas warred against the gods and was condemned to carry the heavens on his shoulders for eternity. He is usually, erroneously, depicted as carrying not the heavens but the world.

The Hyades (the name means 'rain' because in ancient Greece their rising and setting occurred during the wet seasons) are the nearest star cluster to our solar system. They are quite old and the stars fairly dispersed so are best seen through binoculars.

Now continue onwards along the path from Orion's Belt, past Aldebaran, and you will see a compact cluster of stars which sparkle like a casket of diamonds. These are the Pleiades. Most people can see six or seven stars, and binoculars will bring dozens more into view (see figure 12). The entire cluster contains several hundred stars, the brightest of which are 1,200 times brighter than the sun.

12 The Pleiades – and their parents.

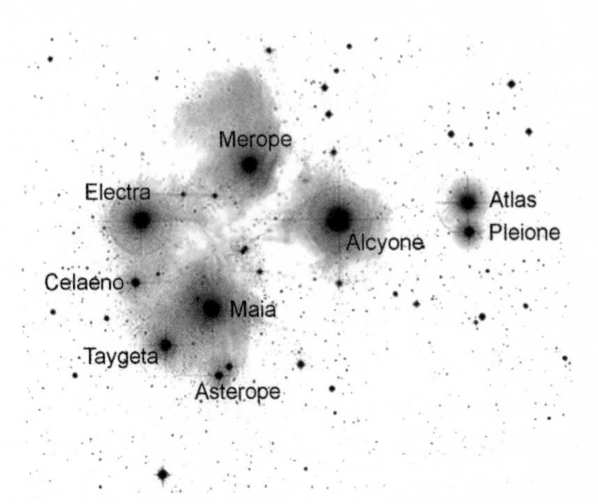

As already noted, cluster stars tend to be young and the Pleiades, with an age of just 50 million years, are infants as far as stars go. The cluster, 410 light-years away, is immersed in a cosmic cloud of dust and ice crystals that reflects the blue light of the brighter stars. The wispy blue nebulosity looks fabulous in long-exposure photographs but is difficult to see visually without a large telescope.

In Greek mythology the sisters are seven nymphs, daughters of the Titans Atlas and Pleione. Actually I'm rather fond of Pleione: she was the very first star I studied. And after many a night in the cold she finally revealed her light curves. Pleione is white-hot and spins on her axis so fast she has the shape of a lens. As she spins she hurls off rings and shells of glowing plasma at irregular intervals. This causes her to fluctuate unpredictably in brightness. Within hours she can rival Atlas by doubling in brightness. This is no mean feat for a star that is, on average, 600 times brighter than the sun. Her fluctuations and eruptions can be observed with a pair of binoculars.

Historically the Pleiades are very significant. Five thousand years ago their rising marked the northern spring equinox; this was known as the Pleiad-month and signified the year's beginning. Their rising and setting divided the year into seasons for sowing and harvest for the Egyptians, the Greeks and the peoples of Asia. Three thousand years ago the Greeks saw the heliacal rising of the Pleiades as the signal to open the sea-lanes. Because their economy depended upon sea trade (they were also pirates), this was perhaps the most important event of the year. The Pleiades became associated with Athene, the great goddess of wisdom and civilisation, and important Grecian temples such as the Acropolis were orientated to their rising or setting.

The Maori celebration of the Pleiades' rising as the beginning of a new year may date back even further. Five thousand years ago the ancestors of the modern Polynesians were trading with great city-states in what is now Indonesia. It was the tradition in Asia to start the new year with the heliacal rising of the Pleiades. It seems likely that the early Polynesians carried this tradition with them as they voyaged into and settled the Pacific.

Diamonds in the sky

RETURNING TO OUR Orion signpost, if you travel along the line of the three Belt stars in the opposite direction to the Pleiades, you will encounter Sirius, the Dog Star. A path from Bellatrix past Betelgeuse will take you to Procyon, the Lesser Dog Star. Together, the first magnitude stars Betelgeuse in Orion, Sirius in Canis Major, and Procyon in Canis Minor form the Summer Triangle. Remember, also, that the two Dog Stars sit on either side of the Milky Way, the celestial Nile.

Unlike most bright stars, the Dogs are neighbours of the sun: Sirius is 8.6 light-years away and Procyon 11.4.

Sirius is the brightest star in our sky, outshone only by the planets Venus, Jupiter, and sometimes Mars. Maori know Sirius as Takurua, the Frost Star. If it shimmers and twinkles furiously when it rises, there will be a heavy frost.

In real terms Sirius is the brightest of the more than 150 stars within a radius of 25 light-years from Earth. In visible light it is 23.5 times brighter than the sun, but because of its higher surface temperature it emits so much ultra-violet radiation its total energy output is equivalent to 32 suns. Procyon, the Lesser Dog, is the eighth brightest star in the sky. It is white-hot, seven times brighter than the sun.

Hidden within the glare of each of these two stars are strange and bizarre objects that, when discovered, hit the headlines of newspapers around the world. The sun and all the other stars are travelling through space at great speed, on vast journeys around the Milky Way. However they are so far away it would take thousands of years for their positions in the sky to change to the degree this could be noticed by the unaided eye. Instead the stars seem to us to form fixed patterns that remain unchanged for generation after generation.

If we could bring to our time someone who lived 5,000 years ago, he or she would find the present star patterns familiar. However if we went a lot further back – to 100,000 years ago, for example – a visitor from that time would find our night sky unrecognisable. Using

13 The changing pattern of the Southern Cross and the Pointers, due to the motion of the stars through space.

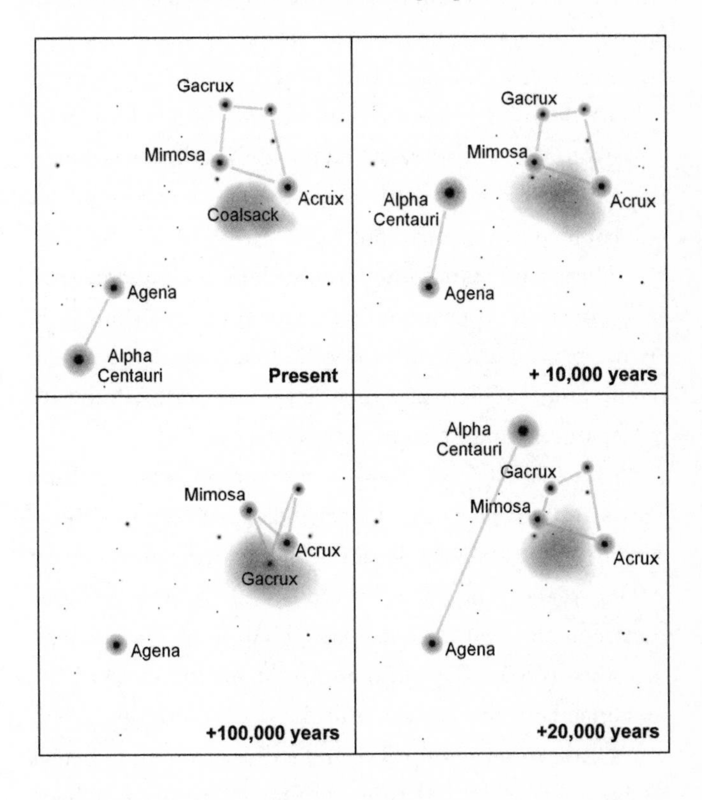

powerful telescopes, today's astronomers can detect and measure the motions of the stars. Figure 13 shows how the pattern of stars which make up the Southern Cross and the Pointers, the two bright stars that follow the Cross around the sky, will change over the next 100,000 years.

In 1834 a German astronomer, Friedrich Bessel, noticed that the motion of Sirius was variable. Instead of moving in a straight line, the star wove a sinuous path through space. Six years later Bessel discovered a similar phenomenon in the motion of the other Dog Star, Procyon. By 1844 he was convinced that the variation in motion was due in both cases to the attraction of an invisible 'dark' companion.

These dark stars weren't really dark at all and in 1862 an American optician and telescope maker, Alvan Clark, while testing a new telescope he had designed, saw the companion of Sirius. Much later, in 1896, Procyon's companion would also be discovered.

The discovery of Sirius's companion was headline news around the world. A new and exotic celestial object had been unmasked. Called the Pup, it orbits Sirius every 50.09 years. The orbit is elongated, so the distance between the two is continually changing. The average distance is 3.3 billion kilometres, about the same as the distance between the sun and the planet Uranus.

The reason the Pup was, and still is, difficult to detect is because it is 9,100 times fainter than Sirius. I have managed to see it only once, and that was in 1976 when the stars were near to maximum separation. When at their closest the two stars cannot be resolved with the largest telescopes on Earth.

What was exotic about the Pup at the time of its discovery was not just that it was faint, but that it was also very hot. In fact it was almost as hot as Sirius itself. A very hot object that is faint must have a very small radiating surface. It turns out that the Pup is not much larger than the Earth.

Further, not only is it small and hot, it is also extraordinarily massive. Sirius has a mass 2.35 times that of the sun. The tiny Pup, with a volume more than a million times less than its big brother, has a mass 99 percent that of the sun. Imagine the entire mass of our solar system compressed into an object just a little larger than the Earth. The Pup is so incredibly dense it exceeds the density of lead by a factor of thousands. A teaspoonful of its matter would weigh a hundred tons.

The companion of Sirius was the first, and nearest, of a new class of star to be discovered – a class now called 'white dwarfs'. Procyon's companion is also a white dwarf and has an orbital period of 40.65 years. White dwarfs are stellar corpses, the fading embers of once brilliant stars. When a star's nuclear fires – the reactions that cause it to emit energy – are exhausted, the star collapses and dies under the crushing force of its own gravity. What remains is a super-dense, white-hot cinder that will, in the aeons to come, cool and become cold and dark.

Even now the surface of the Pup is solidifying. In time the entire star will crystallise into a solid object. Because it is made mostly of carbon, it will become a gigantic crystal, a solid diamond the size of a planet. But woe betide any prospectors who attempt to land on this diamond in the sky. With so much mass concentrated into a small volume, the surface gravity will be enormous. At its surface the average human would weigh 2,000 tonnes. Upon touchdown, spacecraft and astronauts would be instantly flattened and spread over an area the size of a football field.

Jocelyn's lighthouse

WHEN I BEGAN writing this book I decided that I would stick to objects that could be identified with the naked eye. However, because they are so interesting, I have allowed myself two exceptions. Jocelyn's lighthouse is the first.

In 1054 Chinese astronomers recorded a brilliant star having appeared in the constellation of Taurus. There is some evidence the event was also recorded by native Americans. The star was so bright that for a few months it could be seen in broad daylight. It then slowly faded from view. The Chinese called it a new star, but this was not the birth of a star. In the depths of interstellar space a giant star had exploded, a truly awesome event which astronomers call a supernova. For a brief period the star blazed out with the light of a 100 million suns. Nearby worlds were vaporised, and shock waves of atomic particles were propagated for hundreds of light-years in all directions.

From our southern perspective the star would have appeared in the sky below the constellation of Orion, close to the star Zeta Tauri (see figure 11). At this point large telescopes now reveal a glowing mass, which we call the Crab nebula. This is the wreck of the star that exploded 950 years ago. The debris from the explosion is spread over an area five light-years in diameter, and is still moving outwards at the rate of 1,000 kilometres a second. On a clear dark night, well away from city lights, the Crab can just be glimpsed with a pair of binoculars.

In 1968 Jocelyn Bell, a PhD student at Cambridge University in England, discovered a strange pulsating radio source coming from deep space. The radio pulses were so regular and seemed so unnatural that the source was initially dubbed LGM1, little green men. But these weren't signals from aliens trying to contact us; the pulses came from a hitherto unknown object called a pulsar, an object even more bizarre than a white dwarf. Since that initial discovery many other pulsars have been discovered, one of the nearest and most powerful being located near the centre of the Crab nebula.

What had happened? It seemed that when the giant star had exploded it had not been completely destroyed. Its collapsed core was, and is, still there, hissing and pulsating radiation as it spins on its axis 30 times a second. This remnant, the Crab pulsar, contains more than twice the mass of the entire solar system, crushed into a tiny sphere just 18 kilometres in diameter. Its density is so great that a tablespoon of its substance would weigh as much as the Earth.

Under the powerful force of gravity, the pulsar is so dense that not even atoms can exist. Its interior consists of a fluid of sub-atomic particles called neutrons, encased in a solid shell of iron and neutrons. For this reason it is also known as a neutron star.

Essentially it is a gigantic atomic particle the size of a city. Star quakes at its surface form cracks and fissures. From these, a flood of atomic particles spirals out into space, generating a beam of radio waves and light. As the star spins on its axis, the beam sweeps around like the beacon of a lighthouse. The Earth happens to be in the sweep of this lighthouse so we get a pulse of light and radio waves 30 times a second.

As an interesting aside, Jocelyn Bell, who had made this important discovery, did not receive due credit for it. Instead, in 1974 the Nobel Prize for Physics was jointly awarded to her supervisor, Anthony Hewish, and another Cambridge astrophysicist, Martin Ryle, with Hewish cited 'for his decisive role in the discovery of pulsars'.

The event caused considerable controversy, with Sir Fred Hoyle, the most eminent astronomer of the day, moved to make a public protest.

White dwarfs are the burnt-out cinders of ordinary stars; neutron stars are the remains of giant stars. These stellar deaths are significant to our very existence on Earth. When the universe formed it consisted almost entirely of hydrogen and helium. Heavier elements, such as carbon and oxygen, were forged in the interiors of stars. Right now our sun is producing helium; towards the end of its life it will produce carbon.

When stars die they eject much of their substance back into space, sometimes violently in a supernova explosion. With the passage of time the universe has become enriched with heavier elements from which planets, rocks, trees and people are made. Common stars produce the common elements. Rare elements, however, tend to be produced by rare events such as a supernova explosion. It is sobering to think that the gold in the ring on your finger is a lump of star-stuff, forged in a supernova explosion millions of years before the Earth was born.

In time the debris of the Crab nebula will be swept up in the formation of new stars. Like the phoenix, stars rise from their own ashes.

The Southern Cross

FOR NORTHERN navigators the most celebrated object in the sky is Polaris, the North Star. Polaris is a bright second-magnitude star that lies within one degree of the north celestial pole. Its direction gives true north.

When early sailors crossed the equator journeying southward, they noticed that Polaris was lost from view. A new polar marker was required, but unfortunately there was no bright star marking the position of the south celestial pole. To find it they used the Southern Cross, the constellation known officially as Crux. The Southern Cross is the smallest constellation in the sky. Originally it was part of the constellation Centaurus, but because of its importance as a navigational beacon, sixteenth century European explorers and astronomers turned it into a separate constellation. To Maori, who used it as a navigational beacon long before Europeans did, it is Te Punga, the anchor of the Great Waka.

During summer evenings the Milky Way arches across the heavens from north to south. Starting from Orion, cast your eyes along the Milky Way towards and then past Sirius. You are heading south and will eventually encounter the Southern Cross, which is embedded in the Milky Way. In summer the cross lies on its side.

14 Using the Southern Cross to find due south and your latitude.

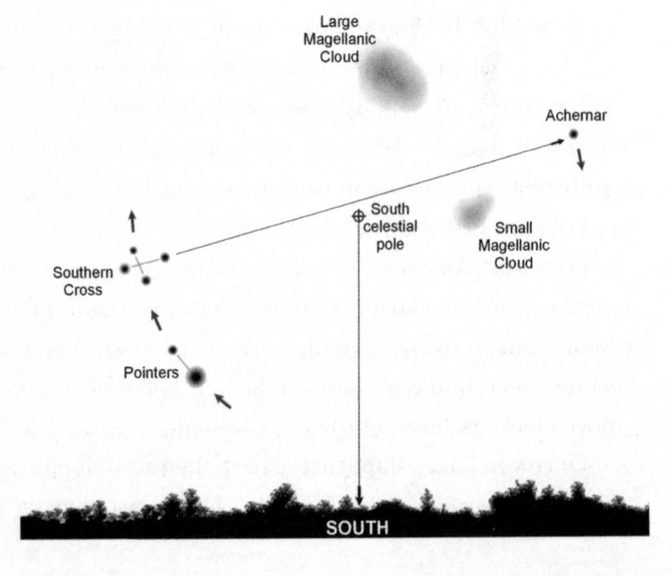

There are several crosses in the southern Milky Way. To identify the real Southern Cross, look for the two bright Pointer stars (see figure 14). Another clue to look for is the Coalsack, a dark cosmic cloud on the edge of the cross. From a dark sky site the Coalsack nebula is very conspicuous because it is darker than the normal night sky and is silhouetted against a brilliant region of the Milky Way. Because the nebula blots out the light of more distant stars, the unaided eye can see only one star in the Coalsack.

To use the Southern Cross as a navigational beacon, think of it as a large arrowhead in the sky with the brightest star, Acrux, at the tip. Now, in your mind draw a straight line through the long axis of the cross, through the tip and across the sky. You will find the arrow is pointing at another bright star. You cannot fail to identify this star, Achernar, because it is the only first-magnitude star in that part of the sky. Approximately halfway along the imaginary line between the cross and Achernar is the south celestial pole. Drop from this point to the horizon and you are looking due south.

Seen from Aotearoa-New Zealand, Crux and Achernar are circumpolar stars – that is, they never set. Their orientation in the sky changes from night to night and from season to season; like hands on a great clock, they move clockwise around the celestial pole.

Once you have established where the celestial pole is, you can determine your latitude. The celestial pole is always a number of degrees above the horizon equal to the latitude at which you are standing. For example, if you measure the angular distance between the south celestial pole and the sea horizon at Wellington, you find it is 41 degrees. That's the latitude of Wellington. Surprisingly, you don't need complex equipment to make this measurement. Polynesian navigators used a notched stick held at arm's length, or sometimes nothing more than the outstretched hand and fingers (see figure 15).

15 Using the outstretched hand to measure angular distances.

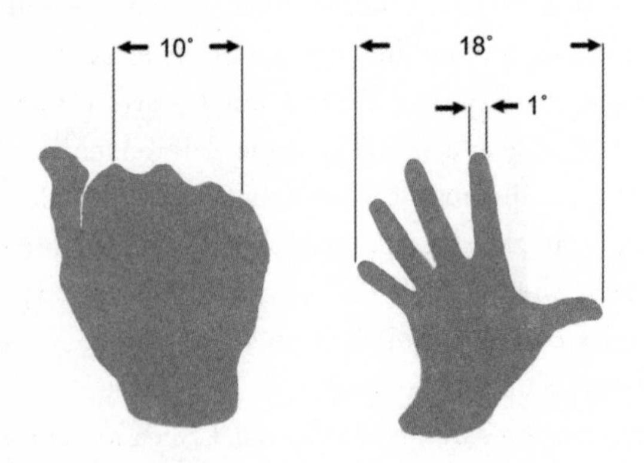

Looking for Goldilocks

THE FIRST THING most visitors from the Northern Hemisphere want to see if they are interested in the stars is the Southern Cross. But for me, when I arrived in New Zealand in 1973, it was the brighter of the two Pointer stars, Alpha Centauri. Alpha Centauri, like the Lost City of Atlantis, is a place of mystery and imagination. But first, a little background.

The Earth is the jewel of our solar system, the only place orbiting the sun that is teeming with life. This comes as no surprise when you examine the nature of life. All life on Earth, from microbes to human beings, is based on the carbon atom. It is difficult to conceive of a different form of life, because carbon is the only element that can join with itself and other elements to create large and complex molecules. Silicon has similar properties but is only about a tenth as good.

Carbon molecules break down at temperatures above 100° Celsius, which is why we use heat for sterilisation. At the other end of the scale, low temperatures slow metabolic functions, and these functions virtually cease at temperatures below 0°C, which is why we preserve organic material by freezing it. To support life, then, the environment must be neither too hot nor too cold. Like the porridge that Goldilocks chose, it must be just right. The Earth provides these conditions because it is just the right distance from the sun. No other planet can occupy this position in our solar system.

Of course if warmth were the only factor, the moon would also support life. But it doesn't, it's barren and lifeless. This is because it doesn't contain another vital ingredient, liquid water. Water is the medium in which the chemical processes of life take place. And the problem as far as the moon is concerned is its lack of atmosphere.

We usually say water boils at 100°C, but this is true only at sea level. Water's boiling point depends upon atmospheric pressure: the thinner the air, the lower the boiling point. You cannot have a hot cup of tea on the top of Mount Everest, because the water would boil well below 100°C. Conversely, in dry valleys below sea level, where the air is denser, water must be heated to well above 100°C before it boils.

The density of a planet's atmosphere depends largely on the planet's mass. Mars, for example, is much smaller than the Earth, and during its summer the air is so thin water boils (meaning it reaches its maximum temperature) at about 3°C. You couldn't have a nice bath on Mars: its hottest water is as cold as our Arctic ocean.

During a Martian winter the air is even thinner, and water boils at 0°C. Liquid water can't exist because, as the temperature rises above zero, ice turns directly into vapour. A glass of water placed on the surface of Mars at this time would literally explode. This is why space travellers need to wear pressurised suits. The human body is mostly water, and in the vacuum of space, without protection, their blood would boil in their veins.

The bigger the planet, the more air and liquid water it can retain at its surface. A planet twice the mass of the Earth would probably have little or no landmass. Instead its entire surface would be covered by water. A planet much more than three times the mass of the Earth would have a dense, suffocating atmosphere of helium and hydrogen, which would be unlikely to support advanced forms of life.

Small planets tend to have little or no atmosphere, while giant planets are blanketed in dense atmospheres. So being the right size is also important. Venus is the right size, but it is too close to the sun and its environment is an inferno. The moon is the right distance from the sun, but it is too small to have an atmosphere. If we are to find other goldilocks worlds – worlds that, like the Earth, support a complex biology – we have to look beyond our solar system to the realms of other stars.

A retinue of planets probably accompanies most, if not all, stars. If only one in a million of these stars has a goldilocks, there will be more than a hundred thousand worlds like ours in the Milky Way. Because each will have evolved its own unique flora and fauna, these places will be magical. Unfortunately, however, they must remain for now places of imagination: the stars are so distant that with current technology we cannot detect, let alone see, planets as small as the Earth.

To be a goldilocks, a planet also needs to orbit the right kind of star. Stars vary enormously in their physical properties, and only five percent could host a goldilocks similar to the Earth. It just so happens that not one but two of these stars are to be found on our cosmic doorstep – at Alpha Centauri. Alpha Centauri is the third brightest star in the sky and shines with a yellow-white hue. The other Pointer star, Agena, is blue-white. The Pointers appear to be close to each other but this is an illusion. Agena is a giant star 526 light-years distant, while Alpha Centauri is just 4.4 light-years away. It appears brilliant because of its relative closeness.

Seen in a telescope, Alpha Centauri is, in fact, a glorious double star. The two stars, both of which are similar to our sun, orbit their common centre of gravity every 80 years. Because of their elliptical orbits the distance between the two is continually changing. At their closest they are separated by 90 light-minutes – a little more than the distance between the sun and Saturn. At their farthest they are almost five light-hours apart, further apart than Neptune and the sun.

Both the Alpha Centauri stars could have their own system of planets. To maintain a stable orbit not affected by the gravitational pull of the other sun, each planet would have to stay well within 16 light-minutes, or 300 million kilometres, of their sun. Planets with orbits similar to that of Mercury, Venus, Earth and Mars would be stable. So a goldilocks *could* be found around either of these stars.

What would it be like on a planet with two suns – one that orbited the brighter of the two Alpha stars at the right distance to receive the same amount of light and heat as the Earth? Its primary sun would look almost identical to ours in the sky, but its second sun, even at its closest, would be 100 times smaller and 215 times fainter. And this second sun wouldn't shed much warmth. The heat and light it provided would be less than one percent of the total.

Inhabitants of this world would enjoy a double sunset-sunrise when the two stars were together in the sky. But pity the poor astronomers: for half the year there would be no darkness. As one sun set, the other would rise.

Future astronauts, arriving on this planet of Alpha Centauri, would undoubtedly look towards home. In the night sky they would see the same constellations we see from Earth. There would, however, be a couple of differences. One of the Pointer stars would be missing, and the northern constellation of Cassiopeia would have an additional star – our sun. The Earth, of course, would not be visible; indeed our entire solar system would have shrunk to a speck of starlight. The astronauts would feel utterly alone. If they were to send a message home, it would be almost nine years before they received a reply.

Is there a goldilocks at Alpha Centauri? Probably not, because there is another major factor to consider – time. We tend to think of our planet as a green and pleasant world, but it has been this way only for the last 450 million years, or 10 percent of its history. As the sun ages it slowly increases in brightness, changing the environments of its planets. For much of the Earth's history its environment was hostile to all but microbes. And in the future, as the sun continues to brighten, the Earth will again become inhospitable.

There is, then, a window in time when complex forms of life can exist on the surface of a planet. Even if there is a potential goldilocks at Alpha Centauri, what are the chances its window on life will coincide with ours? It may have supported life billions of years ago, but have since evolved into a lifeless, searing desert. Or it may be going to blossom a billion years from now.

Looking at this from the other side, if aliens were to visit our solar system at a random point in time they would have only a 10 percent chance of arriving when the Earth is flourishing with life. And human civilisation represents only one-millionth of the Earth's history, so they would have only one chance in a million of encountering human beings.

This immensity of both space and time makes me doubt the Earth has ever been visited by aliens. In a galaxy of billions of stars, the chances of two civilisations arising around neighbouring star systems at the same time is remote. If they are not neighbours, a chance encounter is even more unlikely. If the nearest alien civilisation were a thousand light-years away, our sun wouldn't even be visible without a very large telescope. In addition, these aliens would find a million closer systems to explore before they stumbled upon us.

If they were scanning with radio telescopes, there would be nothing of interest – they would see our solar system as it was a thousand years ago. We weren't broadcasting radio waves then. This may of course change. We have been broadcasting radio waves for over 80 years and these have now travelled some 80 light-years into space. If our nearest technological neighbours are 100 light-years away, in about 20 years' time they will become aware of our existence as our first radio transmissions arrive.

What then of UFOs? Interestingly astronomers, both professional and amateur, who watch the sky often and are conversant with sky phenomena report very few UFOs. Sightings, once identified, are usually found to be common objects, such as the night lights of a distant helicopter or reconnaissance aircraft, or a planet or bright star near the horizon. A bright star will sometimes twinkle wildly and flash different colours. Every year at the Carter Observatory we get UFO reports when the bright star Canopus is near the sea horizon.

Some of the more impressive UFOs turn out to be unusual atmospheric phenomena, such as lenticular clouds or mirages. Lenticular clouds form over mountains and I often see them over the Tararuas. They have a perfect saucer, or lens, shape, and because of their high altitude can be illuminated by the sun after dark.

Once when I was living in Onerahi in Northland I was talking to a friend on the phone, and looking out the window towards the western hills beyond Whangarei, where my friend lived. The sun had set but the sky was still blue.

Suddenly I saw a brilliant object rise up above the hills and move rapidly across the sky. It was cigar-shaped, metallic and shimmering with colours. I quickly told my friend I had to have a closer look. I dropped the phone and found a pair of binoculars, which resolved the mystery. Speeding across the sky was a military aircraft, too high to be seen with the unaided eye. What *could* be seen was its exhaust and vapour trail which, due to its high altitude, was still catching the sunlight.

The following day the local newspaper reported a UFO sighting. A bus full of passengers had stopped to view the unidentified visitor from outer space. Had it not been for my binoculars, until this day I would have been wondering what it was.

If and when we encounter aliens, I expect the experience to be a little more dramatic than lights in the sky, or something that looks like an out-of-focus dustbin lid. UFO reports are simply not fantastic enough. The same applies to 'encounters of the third kind', where people report seeing or being abducted by aliens. What is remarkable about these stories is their lack of imagination. The descriptions of the aliens and their spacecraft invariably sound like they have come out of a cheap science fiction comic or B-grade horror movie.

Don't get me wrong. I think it more than likely that our galaxy is teeming with life, and that thousands of worlds may have evolved intelligent creatures and technological civilisations. I just don't think there is any real evidence the Earth has ever been visited by them.

Nonetheless, I still look at Alpha Centauri and wonder.

And with an

awful, dreadful list

Towards other

galaxies unknown

Ponderously turns

the Milky Way

Boris Pasternak

The Clouds of Magellan

LET'S RETURN NOW to the imaginary line that we drew between the Southern Cross and Achernar to find the south celestial pole (figure 14). On either side of this line you will see two glowing clouds of light. These are the Clouds of Magellan, two small galaxies that are satellites of the Milky Way. Visible only from the Southern Hemisphere, they are the closest galaxies to Earth, and the most distant objects you can easily see with the unaided eye. They are named for the Portuguese explorer Ferdinand Magellan, who first described them in 1519 while on his epic circumnavigation of the globe.

The Large Cloud is 169,000 light-years away and contains about ten billion stars, dark clouds, and glowing nebulae sown with star clusters. In 1987 a star exploded in this galaxy, shining with the light of a hundred million suns. This exploding star, or supernova, was the first that had been visible to the naked eye since 1604, and was co-discovered by New Zealander Albert Jones. Albert, an amateur astronomer who works with a home-made telescope in the backyard of his house in Nelson, is the world's most prolific observer of variable stars (stars that vary in brightness). After his discovery, observatories around the world were alerted and astronomers were able to study the rare event. Although Albert saw the exploding star in 1987, due to its vast distance the explosion had actually occurred 169,000 years before.

The Small Magellanic Cloud is even more distant, 190,000 light-years away. Its tadpole shape is due to the gravitational tidal effects of the Milky Way galaxy. Big galaxies such as the Milky Way grow by cannibalising smaller galaxies; like giant amoeba they move through space, mopping up their smaller brethren. In fact both the Large and Small Clouds are trapped by the gravitational pull and will, with the passage of time, be assimilated into the Milky Way galaxy.

In the constellation of Centaurus, there is a huge, majestic spherical system of stars that can be seen with the naked eye as a fuzzy star (see the star charts at the end of the book). This object, Omega Centauri, contains about five million stars and was originally thought to be the largest globular star cluster in the Milky Way. (Globular clusters are large spherical groups of ancient stars orbiting in the halo around the nucleus of a galaxy.) Recent research, however, suggests that Omega Centauri is the core of another galaxy that was swallowed by the Milky Way a billion years ago.

Close to the Small Magellanic Cloud is a ball of light that looks like an out-of-focus star. If you look through binoculars you will notice its brightness condenses towards the centre. Viewed through a telescope this object, a large globular star cluster called 47 Tucanae, is one of the most magnificent of celestial wonders. It is 140 light-years in diameter and contains perhaps a million stars. The stars crowd towards the centre, where their individual images fuse into a brilliant blaze.

Imagine what the night sky would look like from a world located near the centre of 47 Tucanae, where the stars are 50 to 100 times closer together than in our region of space. Hundreds of stars would be bright enough to be seen in broad daylight. This vast and ancient city of stars drifts in the emptiness of space above and beyond the spiral arms of our Milky Way. From its vantage point, 15,000 light-years from home, you would be able to look out and across the great cosmic whirlpool of stars that is our galaxy.

But despite its grandeur, 47 Tucanae is in fact a stellar graveyard. The bright stars we see are old stars nearing their end. Floating between them are vast numbers of faint white dwarfs and neutron stars – the burnt-out cinders of once brilliant stars. 47 Tucanae is a relic. More than ten billion years old, it is one of the oldest objects in the galaxy.

Shooting stars and ghosts

IF YOU SIT FOR a while beneath the stars, it is more than likely you will see a meteor or 'shooting star' – a streak of light that flashes across the sky. Sometimes these have long trains, or tails. Sometimes they have spectacular colours. And sometimes they are brilliant fireballs that light up the sky.

Despite their nickname, meteors have nothing to do with stars. They are merely tiny pieces of cosmic debris burning up as they enter the Earth's atmosphere. As they pass through the atmosphere, friction heats these tiny fragments to incandescence, and we see a meteor. Most meteors are no larger than a grain of sand. A meteor the size of your fist would momentarily turn the night sky into broad daylight. If the meteor survives passage through the atmosphere and lands on the ground, we call it a meteorite.

As the Earth moves along in its orbit it sweeps up any particles in its path. After midnight our view out into space is in the direction the Earth is moving. Consequently this is when we see most meteors (see figure 16). Meteors we see in the early evening are striking the Earth from behind, and to do this they must be moving faster than the Earth. Hence they are rarer.

Meteors have three sources. Many are rock or iron fragments that have been produced by collisions between asteroids. Asteroids, also called minor planets, are small rocky bodies ranging in size from small boulders to a thousand kilometres in diameter. There are thousands of asteroids in the solar system; most move around the sun between the orbits of Mars and Jupiter. A small number of meteors are fragments from other planets, thrown into space by the violent impact of an asteroid or comet. These two groups of meteors tend to be sporadic: their time of arrival and path through the sky are unpredictable.

The third source of meteors is comets. A bright comet looks like a large fuzzy star with a long ghostly tail. Comets do not flash across the sky; like planets they move sedately among the stars. The core of a comet is a small body, a few kilometres in diameter, composed of porous,

16 The Earth moving through space: relative to the planet's motion, an observer (denoted by white spots) is at the back of the Earth at sundown and out front just before dawn.

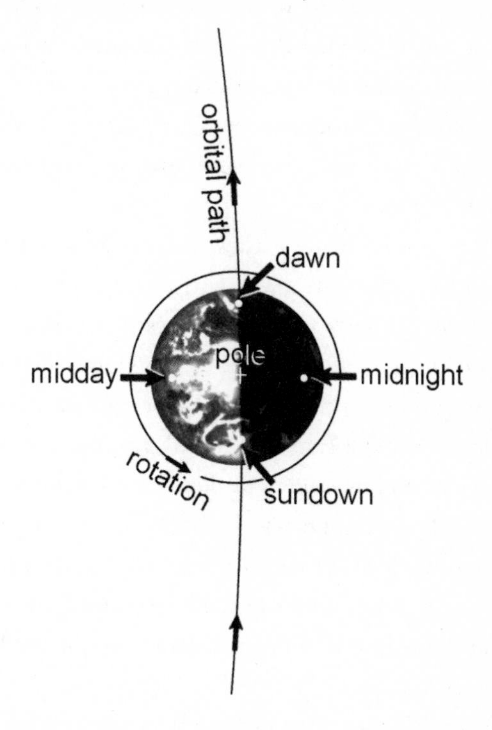

dusty ice. Comets are remnants of the material which formed our solar system, and survive today in the dark outer regions far beyond the orbits of the planets.

Occasionally, due to some disturbance, a comet may fall towards the sun and pass through the inner solar system. When this happens, the sun's radiation heats and evaporates the outer layers, and a glowing halo of dust, ice and gas forms. This halo may swell to tens of thousands of kilometres wide. As well as heat and light, the sun emits a flood of atomic particles, known as the solar wind. This wind drives the comet's halo backwards and a tail forms; in time this tail may grow to millions of kilometres. No matter in which direction a comet is heading, its tail will, of course, always point away from the sun.

Throughout history these mysterious objects have been held in awe. In many early cultures they were regarded as portents of famine, pestilence and war, the death of princes and the fall of kingdoms. These beliefs may originate from the fall of Jerusalem: in 66 AD, shortly before the Romans sacked the city, a comet hung in the sky. As harbingers of doom, comets often became self-fulfilling prophecies. In 1066, for example, when Norman warlords saw a comet in the sky, they took it as evidence of the fall of a kingdom and promptly invaded England, leading to their famous victory at the Battle of Hastings.

When a comet passes through the inner solar system, loose debris is left in its wake and spreads out along its orbit. If the path of this debris intersects the Earth's orbit, a shower of meteors can be seen. A particular shower will occur at the same time each year – when the Earth crosses the old path of the comet – and all the meteors will appear to radiate from the same point in the sky, a position known as the radiant.

17 Comet Halley debris stream: the stream intersects the Earth's orbit at two points, producing meteor showers.

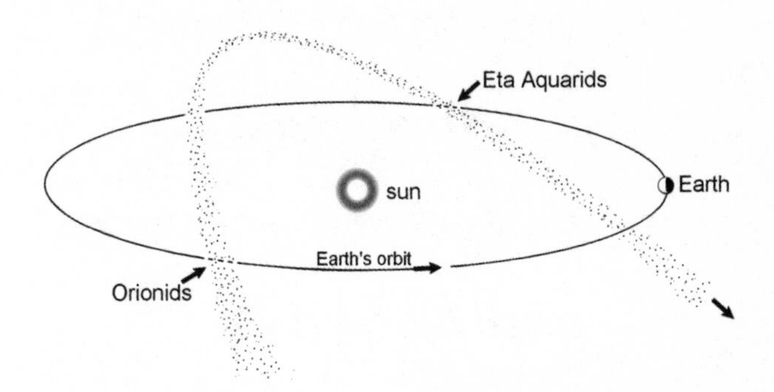

Figure 17 shows the debris stream from Comet Halley. The Earth encounters this stream twice a year. In May it produces the Eta Aquarid meteor shower, and in October the Orionid. A meteor shower is named after the nearest bright star to the radiant, or the constellation in which it occurs. Incidentally, it was Comet Halley that was seen over Jerusalem, and by the Normans.

Upon a slight conjecture I have ventured on a dangerous journey, and I already behold the foothills of new lands. Those who have the courage to continue the search will set foot upon them.

Immanuel Kant,
***Allgemeine Naturgeschichte und Theorie des Himmels* (*Universal Natural History and Theory of Heaven*)**

Hell's huge black spider

TWO OF THE MOST notable constellations in the sky are Orion (The Hunter), whom we have already met, and Scorpius (The Scorpion). Both are set against the Milky Way and composed of bright stars. The two constellations are almost opposite each other in the sky, and have been used around the world and through the ages as seasonal markers. In the Southern Hemisphere Orion dominates the summer evening sky, while the Scorpion is supreme in winter. The two are seen together only in spring and autumn, when one is setting in the west as the other rises in the east.

Both constellations have origins way back in antiquity, probably well before the rise of civilisation. Orion, in all northern cultures, has always been portrayed as a hunter. To the cultures that arose from the great Mesopotamian tradition, Scorpio is a scorpion. But in China it is a dragon. And to the Polynesians it is the fish hook of Maui.

Looking south-west from a point north of the equator, the hook of stars in Scorpius hangs downward below the horizon. It is fishing in the sea. As you sail towards the hook, it slowly rises up. If you continue sailing in that direction, eventually Aotearoa-New Zealand will appear on the horizon below the hook. This is the hook Maui used, in Maori legend, to fish up the islands.

Scorpius is, then, a navigational beacon for sailing from the north to Aotearoa-New Zealand. It is also the country's zenith constellation, and passes directly overhead when you are at the latitude of Aotearoa-New Zealand, around 40° south.

In the Greek story Orion, a giant of a man, was the greatest hunter on Earth. He was a favourite of many of the gods but his boasting upset others. One day he boasted that he would hunt and kill every animal on Earth. This angered Hera, the Queen of the Gods, who sent one of her minions, a monstrous scorpion, to teach him a lesson. The scorpion stung and killed Orion.

Some of the gods, feeling sorry for Orion, immortalised him and placed him in the heavens. But Hera would have none of this; she placed her scorpion at the opposite end. For the rest of eternity the scorpion would chase Orion across the sky. As the scorpion rose, Orion would flee the sky (that is, it would set), only to return when the scorpion entered the underworld (when it, in turn, set).

Scorpius is one of the few constellations that looks something like what it is supposed to be. Antares, a bright

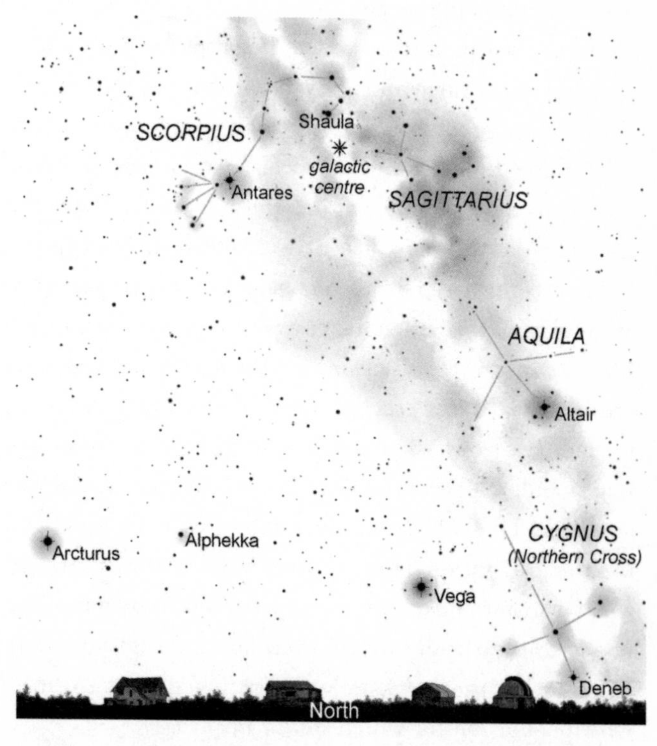

18 The scorpion and galactic centre: Scorpius and the brightest part of the Milky Way pass directly overhead in our winter sky. The direction of galactic centre lies close to the scorpion's tail.

reddish star, marks the heart of the beast and a curve of bright stars forms its tail. The blue-white star Shaula is the sting at the end of the tail. Antares, like Betelgeuse, is a colossal red supergiant star, 400 times the diameter of the sun. This red colossus has a companion star which, due to contrast, appears green. However, this giant ruby and its emerald companion are too close together to be seen without a telescope.

As the scorpion rises we see the largest and most brilliant region of the Milky Way. It appears as a huge bulge, divided by a great dark rift. According to legend this dark rift is the hole in the sky from which the scorpion emerged. And hidden behind these clouds is indeed a monster – at the galactic centre, 27,000 light-years away.

The central region of our galaxy is a vast spherical system of 50 billion stars, crowded together, forming a blaze of light. Like bees around a honey jar, the stars crowd toward the centre until it becomes a cosmic tornado. Here, swathed in a web of gravity, is hell's huge black spider, a gigantic black hole in the fabric of space and time. The hole is a billion kilometres in diameter and has the gravitational pull of three million suns. Snared in this web, stars near the hole are turned into comet-like bodies as they head towards their doom. Eventually they merge with the black hole's accretion disk, a gigantic whirlpool of matter which spirals inwards.

Here is utter oblivion. Anything entering the hole vanishes from the universe. Here even space and time cease to exist. If you were to fall into this hole you would fall through time. You would see the entire future of the universe. But there would be no coming back. These secrets would be for you alone.

Glossary

asteroid Small rocky body orbiting around the sun; also called minor planet. Thousands exist, ranging from small boulders to 1000 kilometres in diameter.

astronomical unit (AU) Mean distance between the Earth and the sun: 149,597,870 kilometres.

binary star Two stars orbiting a common centre of gravity.

black hole Region of space where gravity is so strong nothing can escape, not even light.

blue giant Large, very hot, highly luminous star.

celestial equator Projection of the Earth's equator on to the celestial sphere, dividing the sky into two equal hemispheres.

celestial pole Imaginary projection of the Earth's axis of rotation on to the celestial sphere; a point about which the apparent daily rotation of the stars takes place.

celestial sphere Imaginary sphere surrounding the Earth, on which celestial bodies appear fixed.

comet Small icy body moving in a highly elliptical orbit around the sun. Comets spend most of their time in frozen reaches but periodically their orbits bring them close enough to the sun to be heated, causing formation of a hazy head, and a tail.

conjunction A planet is said to be in conjunction with a star or another planet when it is close to it in the sky. Planets within the Earth's orbit are at inferior conjunction when they are between the Earth and the sun, and superior conjunction when on the far side of the sun. Planets outside of Earth's orbit can only reach superior conjunction.

constellation Recognised patterns of stars, used to divide the sky into 88 regions. The stars of a constellation often have no physical relationship with each other.

cosmology Theories on the origin and nature of the universe.

crescent moon Either a waxing or waning moon, before first or after last quarter.

ecliptic Apparent path of the sun around the celestial sphere produced by the orbital motion of the Earth. This is the plane of our solar system; background stars along the path form the constellations of the zodiac.

elongation Angular distance of a planet from the sun.

equinox Two times during the year when the sun crosses the celestial equator. The southern autumn equinox occurs around 21 March, and the southern spring equinox around 22 September.

evening star Bright planet, usually Venus, seen in the western evening twilight.

fireball Very bright meteor.

first quarter moon The moon has completed a quarter of its cycle, 7 days from new moon, and is half illuminated.

full moon The moon is fully illuminated and has completed one half of its cycle; it is now opposite the sun.

galaxy System of at least one million stars independent of the Milky Way.

heliacal rising Star's rising (becoming visible) just before the sun.

inferior planets Planets that have orbits inside that of the Earth's: Mercury and Venus.

last quarter moon The moon has completed three-quarters of its cycle, 21 days, and is half illuminated.

light speed 299,792 kilometres per second.

light-year Distance travelled by light in one year: 9.4607 million million kilometres.

magnitude Measure of the apparent brightness of a celestial object.

mare Dark patches on the moon, which are large, solidified lava plains.

meridian An imaginary circle on the celestial sphere, which runs through the north and south celestial poles and passes directly overhead. When the sun, moon or a star crosses the meridian, it reaches its highest point in the sky.

meteor Small particle that produces a streak of light as it burns up in the Earth's atmosphere. Often called a shooting star.

meteorite Meteor that survives passage through the atmosphere and reaches the ground.

moon's cycle Time from one new moon to the next: 29.53 days.

morning star Bright planet, usually Venus, seen in the eastern morning twilight.

nebula Cosmic cloud of gas and dust.

neutron star Stellar corpse compressed into a very small, super-dense object made of neutrons.

new moon In modern astronomy: when the moon is between the Earth and the sun and cannot be seen. Ancient: when the moon first appears as a very fine crescent in the western sky.

noon Time at which the sun crosses the meridian.

opposition The position of a planet when exactly opposite the sun in the sky, as seen from the Earth.

phase The apparent change in the shape of the moon (and inferior planets) according to the amount of sunlit hemisphere turned towards the Earth.

planet World orbiting a star. Planets shine only by reflecting the light of their parent star.

precession Cyclic 26,000-year wobble of the Earth on its axis.

pulsar Rapidly rotating neutron star that emits short pulses of radiation at very regular intervals.

retrograde motion Situation where a planet appears to move backwards (westward) against the background stars.

shooting star Meteor.

solstice The moment when the Earth's axis is inclined at its maximum (23.5°) towards the sun. For the south pole this is about 22 December (the southern mid-summer day); for the north pole, about 21 June (southern mid-winter day).

stars Distant suns that appear as points of light.

sun, or **Sun** The star about which the Earth orbits; 'sun' is also used to refer to other stars.

superior planets Planets with orbits outside that of the Earth.

supergiant Colossal, high luminosity star.

supernova Exploding star.

Tropic of Cancer Latitude on Earth at which the sun is directly overhead at the southern winter (northern summer) solstice, 23.5 degrees north.

Tropic of Capricorn Latitude on Earth at which the sun is directly overhead at the southern summer (northern winter) solstice, 23.5 degrees south.

twilight Scattered light from the sun when it is below the horizon: dawn and dusk.

vernal equinox The moment when the sun, moving north, crosses the celestial equator: the northern spring, southern autumn equinox. In many ancient cultures it marked the beginning of the year.

waning moon Last half of the moon's cycle, as the moon moves back towards the sun from full to new.

waxing moon First half of the moon's cycle, as the moon moves away from the sun.

white dwarf Stellar corpse, the small, very dense, collapsed core of a once-brilliant star.

zenith Point on the celestial sphere located directly above the observer at 90 degrees angular distance from the horizon.

zodiac The 12 constellations along the ecliptic: the path of the sun, moon and planets.

Tables and star charts

THE BRIGHTEST STARS

This table lists the 30 brightest stars, in order of brightness, in the entire sky, and the constellation and star chart in which each can be located. The star's 'spectrum' indicates its temperature and colour. The classes, from hottest to coolest, are: O and B – blue-white; A – pure white; F – yellow-white; G – yellow; K – golden yellow to orange; M – orange to red. Each class grades from one to the other. This is denoted by a number following the letter, where 0 is the standard (or pure) colour and 9 is almost that of the next one on the spectrum. For example, a type A0 star is pure white, but a type A5 star is halfway in temperature and colour between class A and F. Our sun is type G2. The star's distance is given in light-years.

Star	Constellation	Star chart	Spectrum	Distance (light-years)
Sirius	Canis Major	Summer	A1	9
Canopus	Carina	Southern	A9	313
Alpha Centauri	Centaurus	Southern	G2	4
Arcturus	Bootes	Autumn, Winter	K1	37
Vega	Lyra	Winter	A0	25
Capella	Auriga	Summer	G6	42
Rigel	Orion	Summer	B8	773
Procyon	Canis Minor	Summer, Autumn	F5	11
Achernar	Eridanus	Southern	B3	144
Betelgeuse	Orion	Summer	M2	522
Agena	Centaurus	Southern	B1	526
Altair	Aquila	Winter, Spring	A7	17
Aldebaran	Taurus	Summer	K5	65
Acrux	Crux	Southern	B1	321
Antares	Scorpius	Winter	M1	604
Spica	Virgo	Autumn, Winter	B1	262
Pollux	Gemini	Summer, Autumn	K0	34
Fomalhaut	Pisces Austrinus	Spring, Southern	A3	25
Mira[1]	Cetus	Spring, Summer	M5	418
Mimosa	Crux	Southern	B0	352
Deneb	Cygnus	Winter	A2	1467
Regulus	Leo	Autumn, Summer	B7	77
Adhara	Canis Major	Summer	B2	431
Castor	Gemini	Summer	A2	52
Gacrux	Crux	Southern	M3	88
Tsih[2]	Cassiopea		B0	613
Shaula	Scorpius	Winter	B1	359
Bellatrix	Orion	Summer	B2	243
Alnath	Taurus	Summer	B7	131
Miaplacidus	Carina	Southern	A1	111

1 Mira varies in brightness over a period of about a year. At its brightest it is a prominent orange-red star. At its faintest it can't be seen without a telescope.

2 Tsih is located in the far north and is not visible from the Southern Hemisphere.

METEOR SHOWERS

This table lists the significant meteor showers visible from the Southern Hemisphere. They are identified on the star charts by an eight-pointed star and a capital letter. The active period of each shower is given, and the date each reaches peak activity.

The Zenith Hourly Rate (ZHR) is the number of meteors per hour that are likely to be seen from a dark sky site when the radiant is at the zenith.

Shower name	Star chart	Shower Activity		ZHR (meteors per hour)
		Period	Max	
Alpha Crucids	South	Jan 6 – Jan 28	Jan 19	5
Alpha Centaurids	South	Jan 28 – Feb 21	Feb 7	25+*
Gamma Normids	South	Feb 25 – Mar 22	Mar 14	8
Beta Pavonids	South	Mar 11 – Apr 16	Apr 7	13
Alpha Scorpids	Winter	Mar 26 – May 12	May 3	10
Pi Puppids	South	Apr 15 – Apr 28	Apr 23	40*
Eta Aquarids	Spring	Apr 19 – May 28	May 3	50
Alpha Capricornids	Winter	Jul 3 – Aug 25	Jul 30	8
Pegasids	Spring	Jul 7 – Jul 11	Jul 10	8
Delta Aquarids South	Spring	Jul 8 – Aug 19	Jul 29	20
Pisces Austrinids	Spring	Jul 9 – Aug 17	Jul 29	8
Delta Aquarids North	Spring	Jul 15 – Aug 25	Aug 12	5
Taurids South	Summer	Sep 15 – Nov 25	Nov 3	10
Orionids	Summer	Oct 2 – Nov 7	Oct 22	25
Alpha Monocerotids	Summer	Nov 15 – Nov 25	Nov 21	5
Leonids	Autumn	Nov 16 – Nov 19	Nov 17	15**
Phoenicids	South	Nov 28 – Dec 9	Dec 6	100*
Geminids	Summer	Dec 13 – Dec 15	Dec 14	120*

* This number of meteors is observed only in certain years; in other years there are fewer.

** This is the normal number of meteors seen per hour, but in certain years (every 33) the numbers can climb into the thousands. These events are called 'meteor storms'.

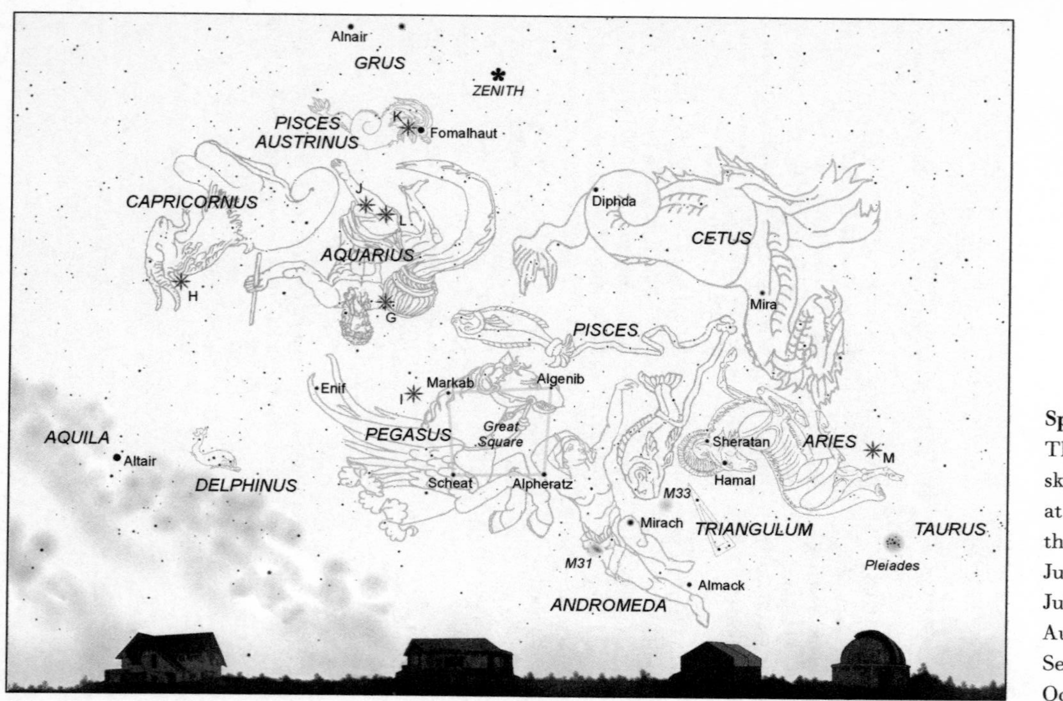

Spring stars

The southern night sky looking due north at the beginning of the month:

June – 5 a.m.
July – 3 a.m.
August – 1 a.m.
September – 11 p.m.
October – 9 p.m.

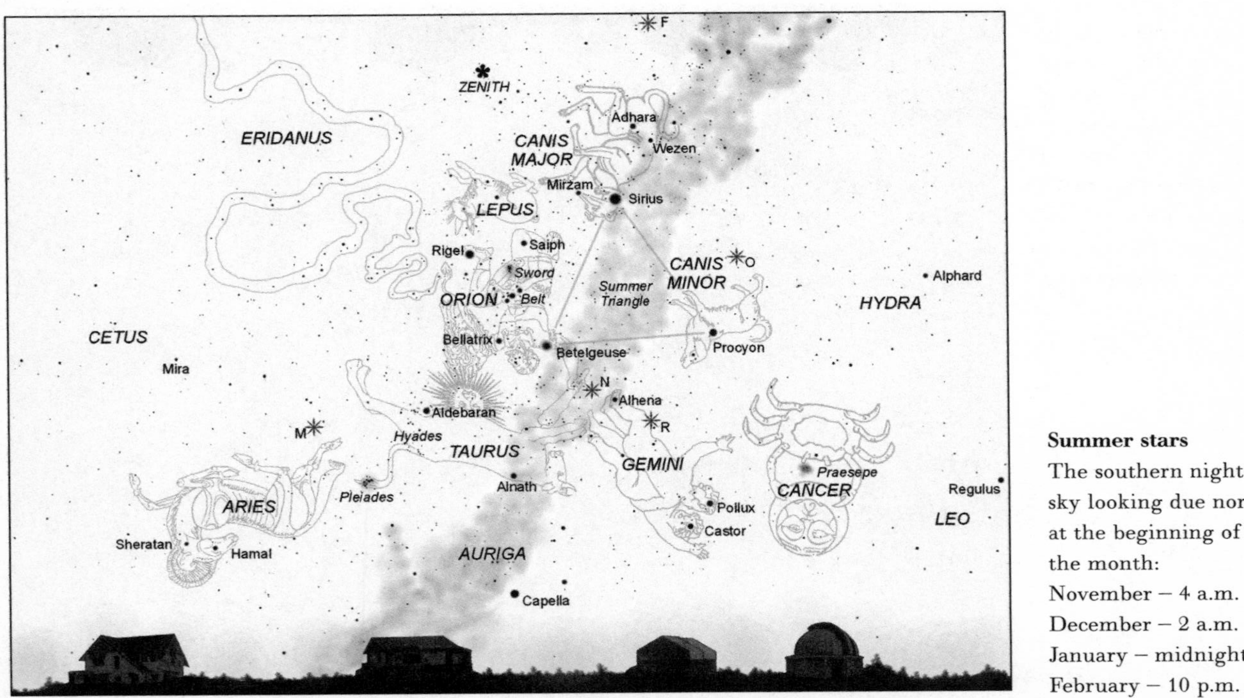

Summer stars

The southern night sky looking due north at the beginning of the month:

November – 4 a.m.
December – 2 a.m.
January – midnight
February – 10 p.m.

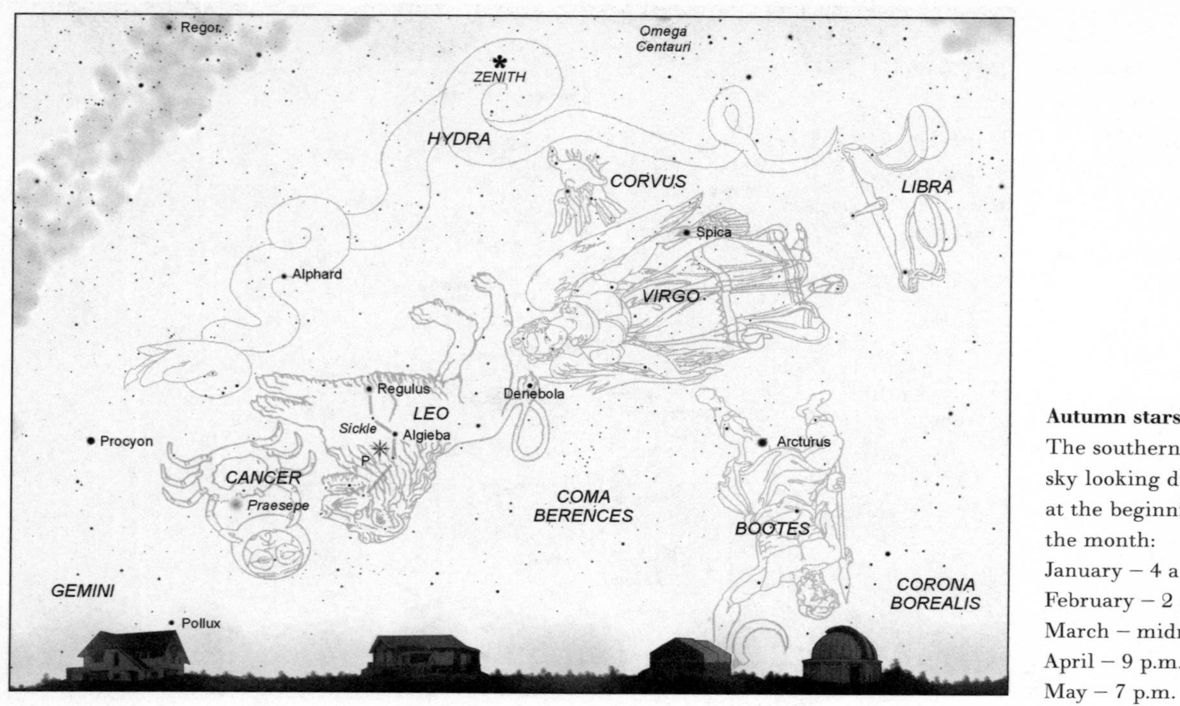

Autumn stars

The southern night sky looking due north at the beginning of the month:

January – 4 a.m.
February – 2 a.m.
March – midnight
April – 9 p.m.
May – 7 p.m.

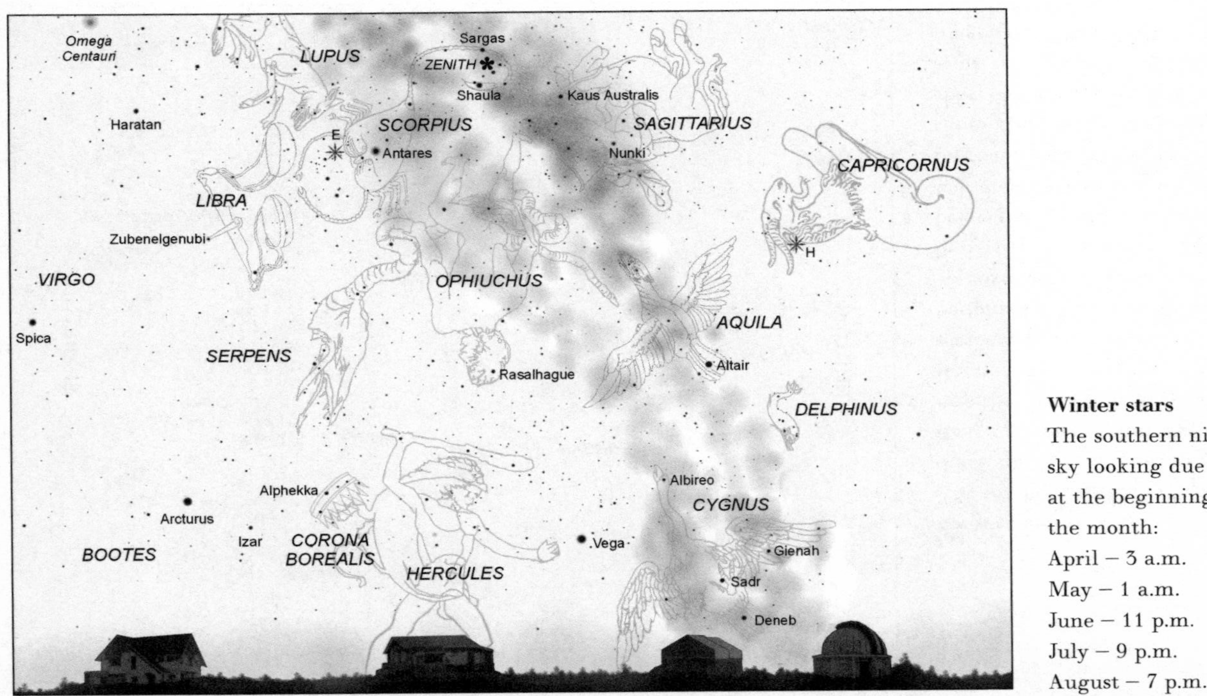

Winter stars

The southern night sky looking due north at the beginning of the month:

April – 3 a.m.
May – 1 a.m.
June – 11 p.m.
July – 9 p.m.
August – 7 p.m.

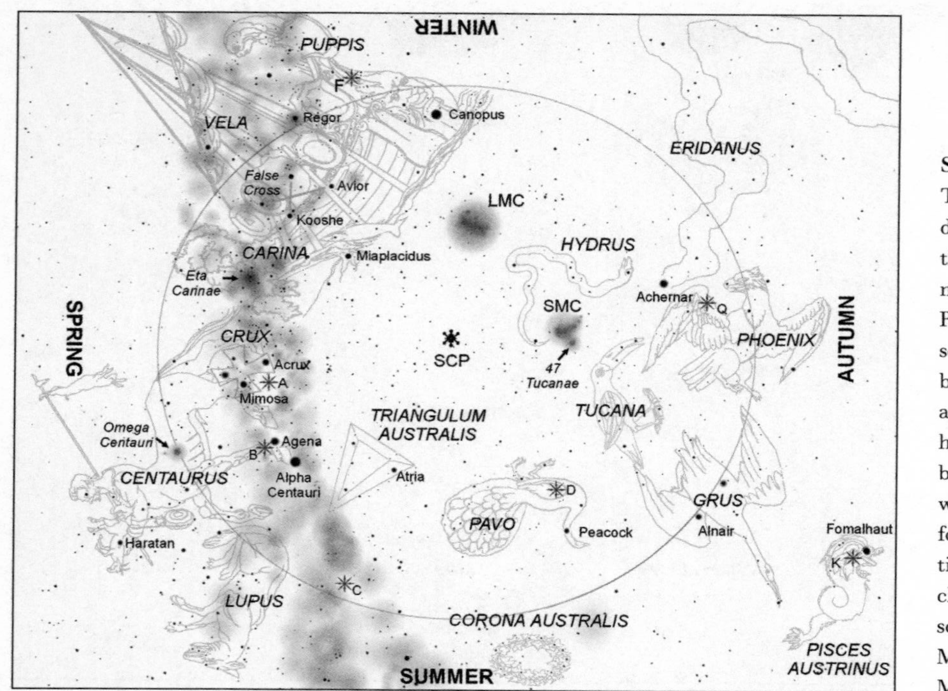

Southern stars
The southern night sky looking due south. The circle encompasses the circumpolar stars – stars that never set from New Zealand. Rotate this chart so the current season is at the bottom. The bottom of the circle will then approximate the position of the horizon. Stars below this point are below the horizon, those above will match the southern night sky for that season (at the dates and times given for the northern sky chart of the same season). SCP: south celestial pole; LMC: Large Magellanic Cloud; SMC: Small Magellanic Cloud.